ENVIRONMENTAL MICROBIOLOGY

ENVIRONMENTAL MICROBIOLOGY

Banwari Lal

CYBER TECH PUBLICATIONS
4264/3, Ansari Road, Daryaganj, New Delhi-110002 (India)
Ph.: 011-23244078. Fax: 011-23280028
E-Mail: cyberpublicationsdelhi@yahoo.com

ENVIRONMENTAL MICROBIOLOGY

Banwari Lal

First Edition 2009

Published by :

G.S. Rawat for **Cyber Tech Publications**
4264/3, Ansari Road, Daryaganj, New Delhi-110002 (India)
Ph.: 011-23244078. Fax: 011-23?80028
E-Mail: cyberpublicationsdelhi@yahoo.com

Printed at : Asian Offset Printers

Preface

"Environmental microbiology " was born at the dawn of the "environmental era" at the beginning of the 1970s. Thirty years of maturation have led to an exciting and vibrant field that has attracted countless numbers of productive and enthusiastic scientists and students at universities, research centers and government agencies around the world."

From the Preface In recent years, the field of environmental microbiology has taken on new importance. But even with a wealth of new research and new interest in the subject, there has never been a single resource to which professionals and students could turn for reliable, detailed coverage of the field. This six-volume set serves as a comprehensive look at the field complete with the latest cutting-edge research.

As we continue to hear about microbial threats to our air, water, and other environmental systems developing the appropriate methodology for their detection, assessing the risk to humans and animals, and preventive measures for avoiding or reducing human exposure to these harmful agents is crucial. Whether you are a microbiologist studying microorganisms and their habitats or working in applied areas such as public health, agriculture, food and water quality, water or wastewater treatment, or resource recovery, this encyclopedia will give you the answers you need.

Undersigned is hopeful enough that this work will be beneficial to students and teachers alike.

— *Author*

CONTENTS

1

Introduction

The study of how microorganisms affect the earth and its atmosphere is called environmental microbiology or microbial ecology.

Environmental microbiology is the study of microbes in the environment and their interactions with each other. Microbes are the tiniest creatures on Earth, yet despite their small size, they have a have a huge impact on us and on our environment.

Most types of microbes remain unknown. It is estimated that we know fewer than 1% of the microbial species on Earth. Yet microbes surround us everywhere — air, water, soil. An average gram of soil contains one billion (1,000,000,000) microbes representing probably several thousand species.

Environmental microbiology is the study of the relationships that exist between microorganisms and the environment.

Environmental microbiology is the relationship of microorganisms with themselves and with their surroundings. Microorganisms have special impact on the whole biosphere. They are the backbone of ecosystems of the zones where light cannot approach. In such zones, chemosynthetic bacteria are present which provide energy and carbon to the other organisms there. Some microbes are decomposers which have ability to recycle the nutrients. So, microbes have a special role in biogeochemical cycles. Microbes, especially bacteria, are of great importance in the sense that their symbiotic relationship (either positive or negative) have special effects on the ecosystem.

Bacteria and their enzymes, along with some fungi and critical nutrient additives are cost effective agents for in-situ remediation of hazardous wastes and subsurface pollution in soils, sediments and wastewaters. The ability of each bacteria strain to degrade toxic waste depends on the nature of each contaminant. Since most sites are typically comprised of multiple pollutant types, the most effective approach to bioremediation is to use a mixture of bacterial species/strains, each specific to the degradation of one or more types of contaminants. It is critical to monitor the composition of the indigenous and added bacterial consortium in order to evaluate the activity level of the bacteria, and to permit modifications of the nutrients and other conditions for optimizing the bioremediation process.

Environmental Microbiology - overview: The pervasive and manifold positive and negative impacts of microbial activities on human and environmental health, and human activities, require that efforts to improve the human condition include efforts to modulate microbial activities. Attainment of this goal will be greatly facilitated by acquisition of a fundamental understanding of microbial functioning in their natural settings, and of parameters that regulate such activities. Our current fragmentary knowledge of microbial processes is based largely on information obtained from individual strains of microbes studied in the laboratory as pure, exponentially-growing cultures in nutrient-rich, homogeneous aqueous solutions.

Sewage treatment is the process that removes the majority of the contaminants from waste-water or sewage and produces both a liquid effluent suitable for disposal to the natural environment and a sludge. To be effective, sewage must be conveyed to a treatment plant by appropriate pipes and infrastructure and the process itself must be subject to regulation and controls.

Sewage is the liquid waste from toilets, baths, showers, kitchens, etc. that is disposed of via sewers. In many areas sewage also includes some liquid waste from industry and commerce. In the UK, the waste from toilets is termed foul waste, the waste from items such as basins, baths, kitchens is termed sullage water, and the industrial and commercial waste is termed trade waste.

The division of household water drains into Greywater and Black water is becoming more common in the United States, with greywater being permitted to be used for watering plants or recycled for flushing toilets. Much sewage also includes some surface water from roofs or hard-standing areas. Municipal wastewater therefore includes residential, commercial, and industrial liquid waste discharges, and may include stormwater runoff.

Sewerage systems that transport liquid waste discharges and stormwater together to a common treatment facility are called combined sewer systems. The construction of combined sewers is a less common practice in the U.S. and Canada than in the past and is no longer accepted within Building Regulations in the UK. Instead, liquid waste and stormwater are collected and conveyed in separate sewer systems, referred to as sanitary sewers and storm sewers in the U.S. and as foul sewers and surface water sewers in the UK. Overflows from foul sewers designed to relieve pressure from heavy rainfall are termed storm sewers or combined sewer overflows.

As rainfall runs over the surface of roofs and the ground, it may pick up various contaminants including soil particles (sediment), metals, organic compounds, animal waste, and oil and grease. Some jurisdictions, such as certain communities located in southern California, require stormwater to receive some level of treatment before being discharged to the environment. Examples of treatment processes used for stormwater include sedimentation basins, wetlands, and vortex separators (to remove coarse solids).

"Conventional wastewater treatment" is the process used in modern wastewater treatment plants that removes the majority of the contaminants from wastewater. This produces a liquid effluent suitable for disposal to the natural environment and also produces a sludge.

The treatment process typically involves the following three stages: Primary treatment Primary treatment is to reduce oils, grease, fats, sand, grit, and coarse (settleable) solids.

Grit removal This stage typically includes a grit channel where the velocity of the incoming wastewater is carefully controlled to allow grit and stones to settle but still maintain all organic material within the flow. Grit and stones need to be removed early in the process to avoid damage to pumps and equipment in the remaining treatment stages.

Screening or Maceration: The grit free liquid is then passed through fixed or rotating screens to remove larger material such as rags. Screenings are collected and may be returned to the sludge treatment plant or may be disposed of off site by landfilling or incineration. Maceration, in which solids are cut into small particles through the use of rotating knife edges mounted on a revolving cylinder, is used in plants that are able to process this particulate waste. Macerators are, however, more expensive to maintain and are less reliable than physical screens.

Sedimentation

Primary sedimentation tank at a rural treatment plantIn almost all plants there is a sedimentation stage where the sewage is allowed to stand in large tanks so that faecal solids can settle and floating material such as grease and plastics can rise to the surface and be skimmed off. The main purpose of the primary stage is to produce a generally homogeneous liquid capable of being treated biologically together with a sludge that can be separately treated or processed. Primary settlement tanks are usually equipped with mechanically driven scrapers that continually drive the collected sludge towards a hopper in the base of the tank from where it can be pumped to further sludge treatment stages.

Secondary treatment Secondary treatment is designed to substantially degrade the biological content of the sewage. The majority of municipal and industrial plants treat the settled sewage liquor using aerobic biological processes. For this to be effective, the biota require both oxygen and a substrate on which to live. There are number of ways in which this is done. In all these methods, the bacteria and protozoa consume biodegradable soluble organic contaminants (e.g. sugars, fats, organic short-chain carbon molecules, etc.) and bind much of the less soluble fractions into floc particles.

Roughing Filters Roughing filters are intended to treat particularly strong or variable organic loads. They are typically tall, columnar filters filled with open synthetic filter media to which sewage is applied at a relatively high rate. The design of the filters allows high hydraulic loading and a high flow-through of air. The resultant liquor is usually within the normal range for conventional treatment processes. Activated sludge Activated sludge plants use a variety of mechanisms and processes to use dissolved oxygen to generate a biological floc that substantially removes organic material. It also traps particulate material and can, under ideal conditions, convert ammonia to nitrite or nitrate.

Filter Beds

Trickling filter bed using plastic mediaIn older plants and plants receiving more variable loads, trickling filter beds are used where the settled sewage liquor is spread onto the surface of a deep bed made up of coke (carbonised coal), rocks or specially fabricated plastic media with high surface areas. The liquor is distributed through perforated rotating arms radiating from a central pivot. The distributed liquor trickles through this bed and is collected in drains at the base. These drains also provide a source of air which percolates up through the bed, keeping it aerobic. Biological film comprising of bacteria, protozoa and fungi forms on all the available surfaces and this provides the required biological treatment capability to effect the reduction in organic content.

Rotating Plates and Spirals In some smaller plants slowly revolving plates or spirals are used which are partially submerged in the liquor. A biotic floc is created which provides the required substrate.

Secondary sedimentation The final step in the secondary treatment stage is to settle out the biological floc or filter material and produce an effluent with very low levels of organic material and suspended matter.

Secondary Sedimentation tank at a rural treatment plantTertiary treatment Tertiary treatment provides a final stage to raise the effluent quality to the standard required before it is discharged to the receiving environment (sea, river, lake, ground, etc.) More than one tertiary treatment process may be used at any treament plant. If disinfection is practiced, it is always the final process.

Effluent polishing Sand filtration Sand filtration removes much of the residual suspended matter.

Lagooning Lagoons provides settlement and further biological improvement through storage in large man-made ponds or lagoons.

Constructed wetlands These include engineered reed beds and a range of similar methodologies, all of which provide a high degree of aerobic biological improvement and can often be used instead of secondary treatment for small communities.

Nutrient removal Wastewater may also contain high levels of nutrients (nitrogen and phosphorus) that in certain forms may be toxic to fish and invertebrates at very low concentrations(e.g. ammonia) or that can create nuisance conditions in the receiving environment (e.g. weed or algal growth). Although the growth of weeds and algae may seem to be primarily an aesthetic issue, algae can produce toxins, and in dying their decay and consumption by bacteria in the environment can result in the depletion of oxygen in the water and the possible consequential suffocation of fish. Where receiving rivers discharge to lakes or shallow seas, the added nutrients can cause severe and sometimes irreversible eutrophication with the loss of many sensitive clean water species. The removal of nitrogen and/or phosphorus from wastewater can be achieved either biologically or by chemical precipitation treatment processes.

Nitrogen removal Biological treatment of nitrogen generally involves creating conditions within the treatment process for bacteria to convert the ammonia to nitrate, and then allowing other bacteria to reducing the nitrate to nitrogen gas, which is released to the atmosphere. Sand filters, lagooning and the use of reed beds can all be used to reduce nitrogen. Sometimes the conversion of toxic ammonia to nitrate alone is referred to as tertiary treatment.

Phosphorous removal The biological treatment of wastewater to remove phosphorus also involves the design and creation of specific environmental conditions within a treatment plant to enable specific bacteria to bio-accumulate large quantities of phosphorus. When the bacteria containing the phosphorus are removed, the resulting bacterial biosolids often have a high fertilizer value. Phosphorus can also be removed by chemical precipitation using (commonly) salts of iron (i.e. ferric chloride) or aluminum (i.e. alum). The resulting chemical sludge, however, is difficult to dispose of, and the use of chemicals in the treatment process is expensive and makes operation difficult and often messy.

Disinfection Disinfecting substantially reduces the numbers of living organisms in the water to be discharged. The effectiveness of disinfection depends on the quality of the water being treated, the type of disinfection being used, the application rate, the contact time and environmental variables.

Turbid water will be treated less successfully since solid matter can shield organisms, especially from Ultraviolet light or if contact times are low. Generally, short contact times, low doses and high flows all mitigate against effective disinfection. Common methods of disinfection include ozone, chlorine, or UV light. Chloramine, which is used for drinking water, is not used in waste water treatment because of its persistence.

Int J Syst Evol Microbiol, 2005 Jan, 55(Pt 1), 433 - 436

Nocardia takedensis sp. nov., isolated from moat sediment and scumming activated sludge; Yamamura H et al.; Chemotaxonomic and morphological characterization of two actinomycete strains, MS1-3(T) and AS4-2, respectively isolated from moat sediment and scumming activated sludge, was carried out. This characterization clearly demonstrated that strains MS1-3(T) and AS4-2 belong to the genus Nocardia. 16S rRNA gene sequencing studies showed that these isolates are most closely related to Nocardia beijingensis (98.1-98.3 % similarity), Nocardia brasiliensis (97.9-98.0 %) and Nocardia tenerifensis (97.8-97.9 %).

However, the results of DNA-DNA hybridizations and physiological and biochemical tests showed that strains MS1-3(T) and AS4-2 could be differentiated from their closest phylogenetic relatives both genotypically and phenotypically. It is proposed that the two isolates be classified as representatives of a novel species of Nocardia, Nocardia takedensis sp. nov. The type strain is MS1-3(T) (=NBRC 100417(T)=DSM 44801(T)); AS4-2 (=NBRC 100418=DSM 44802) is a reference strain.

Int J Syst Evol Microbiol, 2005 Jan, 55(Pt 1), 363 - 368

Shewanella decolorationis sp. nov., a dye-decolorizing bacterium isolated from activated sludge of a waste-water treatment plant; Xu M et al.; A highly efficient dye-decolorizing bacterium, strain S12(T), was isolated from activated sludge of a textile-printing waste-water treatment plant in Guangzhou, China. The cells were Gram-negative and motile by means of a single polar flagellum.

The strain was capable of anaerobic growth either by fermentation of glucose or by anaerobic respiration and utilized a variety of electron acceptors, including nitrate, iron oxide and thiosulfate. The physiological properties, tested by using the Biolog GN2 system, were similar to those of the genus of Shewanella. Analysis of the nearly complete 16S rRNA gene sequence of strain S12(T) showed the highest similarity (98 and 97 %, respectively) to Shewanella baltica and Shewanella putrefaciens.

However, the level of gyrB similarity between strain S12(T) and S. putrefaciens was 87 %. DNA from strain S12(T) showed 41.8 and 41.9 % DNA relatedness, respectively, to the DNA of S. baltica DSM 9439(T) and S. putrefaciens DSM 6067(T). The DNA G+C content of strain S12(T) was 49.3 mol%. The predominant menaquinone was MK-7 and the predominant ubiquinones were Q-7 and Q-8. The dominant fatty acids were 15 : 0, 16 : 0, iso-15 : 0 and 16 : 1omega7c, similar to the profiles of other Shewanella species. On the basis of its physiological and molecular properties, strain S12(T) appears to represent a novel species of the genus Shewanella, for which the name Shewanella decolorationis sp. nov. is proposed. The type strain is S12(T) (=CCTCC M 203093(T)=IAM 15094(T)).

Int J Syst Evol Microbiol, 2005 Jan, 55(Pt 1), 341 - 344

Hydrogenophaga defluvii sp. nov. and Hydrogenophaga atypica sp. nov., isolated from activated sludge; Kampfer P et al.; Two Gram-negative, oxidase-positive rods (strains BSB 9.5(T) and BSB 41.8(T)) isolated from wastewater were studied using a polyphasic approach. 16S rRNA gene sequence comparisons demonstrated that both strains cluster phylogenetically within the family Comamonadaceae: the two strains shared 99.9 % 16S rRNA gene sequence similarity and were most closely related to the type strains of Hydrogenophaga palleronii (98.5 %) and Hydrogenophaga taeniospiralis (98.0 %). The fatty acid patterns and substrate-utilization profiles displayed similarity to the those of the five Hydrogenophaga species with validly published names, although clear differentiating characteristics were also observed. The two strains showed DNA-DNA hybridization values of 51 % with respect to each other. No close similarities to any other Hydrogenophaga species were detected in hybridization experiments with the genomic DNAs. On the basis of these results, two novel Hydrogenophaga species, Hydrogenophaga defluvii sp. nov. and Hydrogenophaga atypica sp. nov. are proposed, with BSB 9.5(T) (=DSM 15341(T)=CIP 108119(T)) and BSB 41.8(T) (=DSM 15342(T)=CIP 108118(T)) as the respective type strains.

FEMS Microbiol Rev, 2005 Jan, 29(1), 49 - 64

"Microthrix parvicella", a filamentous bacterium causing bulking and foaming in activated sludge systems: a review of current knowledge; Rossetti S et al.; This review summarizes the microbiology and physiology of "Microthrix parvicella" and the methods of its growth control in activated sludge wastewater treatment plants. This filamentous bacterium is of high interest because of its worldwide involvement in severe bulking and foaming at wastewater treatment plants. We present a critical analysis of physiological and kinetic data on

"M. parvicella" and discuss its growth and storage abilities in various environments with the aim of understanding the strategies of this organism to successfully compete with other bacteria in activated sludge. Additionally, this review elaborates on research needs for defining reliable control strategies of bulking and foaming based on key features of "M. parvicella".

Environ Toxicol Chem, 2004 Dec, 23(12), 2790 - 8

Effect of ethoxylate number and alkyl chain length on the pathway and kinetics of linear alcohol ethoxylate biodegradation in activated sludge; Itrich NR et al.; Batch activated-sludge die-away studies were conducted with various pure homologs to determine the effect of ethoxylate number and alkyl chain length on the kinetics of primary and ultimate biodegradation of linear alcohol ethoxylates. The 14C-(ethoxylate) homologs C14E1, C14E3, C14E6, and C14E9 were used to investigate the effect of ethoxylate number, and 14C-(ethoxylate) homologs C12E6, C14E6, and C16E6 were used to examine the effect of chain length. Activated sludge was dosed with a trace concentration (0.2 microM) of each homolog, and the disappearance of parent, formation of metabolites, production of 14CO2, and uptake into solids were monitored with time. Ethoxylate number had little effect on the first-order decay rates for primary biodegradation, which ranged from 61 to 78 h(-1). However, alkyl chain length had a larger effect, with the C16 chain-length homolog exhibiting a slower rate of parent decay (18 h(-1)) compared to its corresponding C12 and C14 homologs (61-69 h(-1)). Ethoxylate number affected the mechanism of biodegradation, with fission of the central ether bond to yield the corresponding fatty alcohol and (poly)ethylene glycol group increasing in dominance with increasing ethoxylate number. Based upon the measured rates of primary biodegradation, removal of parent during activated-sludge treatment was predicted to range between 99.7 and 99.8% for all homologs except C16E6, which had a predicted removal of

98.9%. Based upon the measured rates of ultimate biodegradation, removal of ethoxylate-containing metabolites was predicted to exceed 83% for all homologs. These predictions corresponded closely with previously published removal measurements in laboratory continuous activated-sludge systems and actual treatment plants.

Applied Biotechnology

Influence of substrate surface loading on the kinetic behaviour of aerobic granules; Liu Y et al.; In the aerobic granular sludge reactor, the substrate loading is related to the size of the aerobic granules cultivated.

This study investigated the influence of substrate surface loading on the growth and substrate-utilization kinetics of aerobic granules. Results showed that microbial surface growth rate and surface biodegradation rate are fairly related to the substrate surface loading by the Monod-type equation. In this study, both the theoretical maximum growth yield and the Pirt maintenance coefficient were determined.

It was found that the estimated theoretical maximum growth yield of aerobic granules was as low as 0.2 g biomass g(-1) chemical oxygen demand (COD) and 10-40% of input substrate-COD was consumed through the maintenance metabolism, while experimental results further showed that the unit oxygen uptake by aerobic granules was 0.68 g oxygen g(-1) COD, which was much higher than that reported in activated sludge processes. Based on the growth yield and unit oxygen uptake determined, an oxidative assimilation equation of acetate-fed aerobic granules was derived; and this was confirmed by respirometric tests.

In aerobic granular culture, about 74% of the input substrate-carbon was converted to carbon dioxide. The growth yield of aerobic granules was three times lower than that of activated sludge. It is likely that high carbon dioxide production is the main cause of the low growth yield of aerobic granules, indicating a possible energy uncoupling in aerobic granular culture.

Comparing activated sludge and aerobic granules as microbial inocula for phenol biodegradation; Tay ST et al.; Activated sludge and acetate-fed granules were used as microbial inocula to start up two sequencing batch reactors (R1, R2) for phenol biodegradation. The reactors were operated in 4-h cycles at a phenol loading of 1.8 kg m(-3) day(-1). The biomass in R1 failed to remove phenol and completely washed out after 4 days. R2 experienced initial difficulty in removing phenol, but the biomass acclimated quickly and effluent phenol concentrations declined to 0.3 mg l(-1) from day 3. The acetate-fed granules were covered with bacterial rods, but filamentous bacteria with sheaths, presumably to shield against toxicity, quickly emerged as the dominant morphotype upon phenol exposure. Bacterial adaptation to phenol also took the form of modifications in enzyme activity and increased production of extracellular polymers. 16S rRNA gene fingerprints revealed a slight decrease in bacterial diversity from day 0 to day 3 in R1, prior to process failure. In R2, a clear shift in community structure was observed as the seed evolved into phenol-degrading granules without losing species-richness. The results highlight the effectiveness of granules over activated sludge as seed for reactors treating toxic wastewaters.

J Environ Qual, 2005 Jan-Feb, 34(1), 105 - 13

Processes for managing pathogens; Godfree A et al.; Wastewater contains human, animal, and plant pathogens capable of causing viral, bacterial, or parasitic infections. There are several routes whereby sewage pathogens may affect human health, including direct contact, contamination of food crops, zoonoses, and vectors. The range and numbers of pathogens in municipal wastewater vary with the level of endemic disease in the community, discharges from commercial activities, and seasonal factors. Regulations to control pathogen risk in the United States and Europe arising from land application of biosolids are based on the concept of multiple barriers to the prevention of transmission. The barriers are:

(i) treatment to reduce pathogen content and vector attraction,

(ii) restrictions on crops grown on land to which biosolids have been applied, and

(iii) minimum intervals following application and grazing or harvesting.

Wastewater treatment reduces number of pathogens in the wastewater by concentrating them with the solids in the sludge. Although some treatment processes are designed specifically to inactivate pathogens, many are not, and the actual mechanisms of microbial inactivation are not fully understood for all processes. Vector attraction is reduced by stabilization (reduction of readily biodegradable material) and/or incorporation immediately following application. Concerns about health risks have renewed interest in the effects of treatment (on pathogens) and advanced treatment methods, and work performed in the United States suggests that Class A pathogen reduction can be achieved less expensively than previously thought. Effective pathogen risk management requires control to the complete chain of sludge treatment, biosolids handling and application, and post-application activities. This may be achieved by adherence to quality management systems based on hazard analysis critical control point (HACCP) principles.

Evaluation of microscopic techniques (epifluorescence microscopy, CLSM, TPE-LSM) as a basis for the quantitative image analysis of activated sludge; Lopez C et al.; Microscopic techniques ranging from epifluorescence microscopy to confocal laser scanning microscopy (CLSM) and two photon excitation laser scanning microscopy (TPE-LSM) combined with fluorescent stains can help to evaluate complex microbial aggregates such as activated sludge flocs. To determine the application limits of these microscopic techniques, activated sludge samples from three different sources were evaluated after staining with a fluorescent viability indicator (Baclight Bacterial Viability Kit, Molecular Probes). Image analysis routines were developed to quantify overall amounts of red and green stained cells, location of stained cells within the flocs, and the spatial organization in clusters and filaments. It was found that the selection of the appropriate microscopic technique depends strongly on the type of microbial aggregates being analyzed. For flocs with high cell density, the use of TPE-LSM is preferred, since it provides a clearer image of the internal structure of the aggregate. Epifluorescence microscopy did not allow to reliably quantify red stained cells in dense aggregates. CLSM did not adequately image the internal filamentous structure and the location of stained cells within dense flocs. However, for typical activated sludge flocs epifluorescence and CLSM proved adequate.

Parallel hybrid modeling methods for a full-scale cokes wastewater treatment plant; Lee DS et al.; Parallel hybrid modeling methods are applied to a full-scale cokes wastewater treatment plant. Within the hybrid model structure, a mechanistic model specifies the basic dynamics of the relevant process and a non-parametric model compensates for the inaccuracy of the mechanistic model. First, a simplified mechanistic model is developed based on Activated Sludge Model No. 1 and the specific process knowledge of the cokes wastewater treatment process. Then, the mechanistic model is combined with five different non-parametric models - feedforward back-propagation neural network, radial basis function network, linear partial least squares (PLS), quadratic PLS and neural network PLS (NNPLS) - in parallel configuration. These models are identified with the same data obtained from the plant operation to predict dynamic behavior of the process. The performance of each parallel hybrid model is compared based on their ease of model building, prediction accuracy and interpretability. For this application, the parallel hybrid model with NNPLS as non-parametric model gives better performance than other parallel hybrid models. In addition, the NNPLS model is used to analyze the behavior of the operation data in the reduced space and allows for fault detection and isolation.

Recovery of an environmental chlamydia strain from activated sludge by co-cultivation with Acanthamoeba sp; Collingro A et al.; Chlamydiae are a unique group of obligate intracellular bacteria comprising important pathogens of vertebrates as well as symbionts of free-living amoebae. Although there is ample molecular evidence for a huge diversity and wide distribution of chlamydiae in nature, environmental chlamydiae are currently represented by only few isolates. This paper reports the recovery of a novel environmental chlamydia strain from activated sludge by co-cultivation with Acanthamoeba sp. The recovered environmental chlamydia strain UV-7 showed the characteristic morphology of chlamydial developmental stages as revealed by electron microscopy and was identified as a new member of the family Parachlamydiaceae (98.7 % 16S rRNA sequence similarity to Parachlamydia acanthamoebae). Infection studies suggested that Parachlamydia sp. UV-7 is not confined to amoeba hosts but is also able to invade mammalian cells. These findings outline a new straightforward approach to retrieving environmental chlamydiae from nature without prior, tedious isolation and cultivation of their natural host cells, and lend further support to suggested implications of environmental chlamydiae for public health.

Fenton peroxidation improves the drying performance of waste activated sludge; Dewil R et al.; Advanced sludge treatment processes (AST) reduce the amount of sludge produced and improve the dewaterability, thus probably also affecting the heat transfer properties and the drying characteristics of the sludge. This paper studies the influence of the Fenton peroxidation on the thermal conductivity of the sludge. Results demonstrate that the Fenton's peroxidation positively influences the sludge cake consistency and hence enhances the mechanical dewaterability and the drying characteristics of the dewatered sludge. For the two sludges used in this study, i.e. obtained from the wastewater treatment plants (WWTP) of Tienen and Sint-Niklaas - the dry solids content of the mechanically dewatered sludge increased from 22.5% to 40.3% and from 18.7% to 35.2%, respectively.

The effective thermal conductivity k(e) of the untreated and the peroxidized sludges is measured and used to determine the heat transfer coefficient h(s). An average improvement for k(e) of 16.7% (Tienen) and 5.8% (Sint-Niklaas) was observed. Consequently the value of h(s) increased with 15.6% (Tienen) and 5.0% (Sint-Niklaas). This increased heat transfer coefficient in combination with the increased dewaterability has direct implications on the design of sludge dryers. A plate-to-plate calculation of a multiple hearth dryer illustrates that the number of plates required to dry the peroxidized sludge to 90% DS is less than half the number of plates needed to dry untreated sludge. This results in reduced dryer dimensions or a higher capacity for an existing dryer of given dimensions.

Discrepancies in the widely applied GAM42a fluorescence in situ hybridisation probe for Gammaproteobacteria; Siyambalapitiya N et al.; A bacterial culture collection of 104 strains was obtained from an activated sludge wastewater treatment plant to pursue studies into microbial flocculation. Characterisation of the culture collection using a polyphasic approach indicated seven isolates, phylogenetically affiliated with the deep-branching Xanthomonas group of the class Gammaproteobacteria, were unable to hybridise the GAM42a fluorescence in situ hybridisation (FISH) probe for Gammaproteobacteria. The sequence of the GAM42a probe target region in the 23S rRNA gene of these isolates was determined to have mismatches to GAM42a. Probes perfectly targeting the mismatches (GAM42a_T1038_G1031, and GAM42a_T1038 and GAM42a_A1041_A1040) were synthesised, and used in conjunction with GAM42a in FISH to study the Gammaproteobacteria community structure in one full-scale activated sludge plant. Several bacteria in the activated sludge biomass bound the modified probes demonstrating their presence and the fact that these Gammaproteobacteria have been overlooked in community structure analyses of activated sludge.

Effect of copper in the protistan community of activated sludge; Nicolau A et al.; Protists have proved to be an interesting tool for assessing the occurrence of pollution in wastewater treatment systems along with its role in the control of pollution itself through grazing of dispersed bacteria and maintenance of a healthy trophic web in those artificial ecosystems. Two sets of assays were carried on in a bench scale pilot plant in order to study the response of the activated sludge community of protists to the exposure of copper: the first set was carried on with synthetic sewage and the second one with real sewage. The results emphasize the ability of activated sludge biological communities to survive and to react to toxicants and highlight the role of protistan communities as indicators of toxicants entrance in treatment systems.

Toxicity of metals and organic chemicals evaluated with bioluminescence assays; Ren S et al.; The development of a bioluminescent sensor organism (Shk1) that was created for assessing wastewater toxicity was reported several years ago. In order to establish a test battery to better characterize wastewater toxicity, additional luminescent sensor organisms were later created. The present study focused on one promising candidate (PM6), a Pseudomonas spp. strain, because of its high level of luminescence compared to that of other newly created organisms. Using a batch toxicity testing protocol, the toxicity of 7 metals and 25 organic compounds was evaluated with the PM6 and Shk1 assays. Results indicated that the toxicity data of the PM6 and the Shk1 assays were correlated, and no assay appeared to be particularly more sensitive to a group of toxicants than the other assay. The results of the PM6 and Shk1 assays were further evaluated by comparing with the results of the Vibrio fischeri luminescence inhibition assay and activated sludge inhibition assays. Data suggested that PM6 and Shk1 more closely represented activated sludge organisms than V. fischeri. The suitability of using PM6 and Shk1 for assessing wastewater toxicity on activated sludge, both individually and in a test battery, was discussed.

Aerobic biodegradation of organotin compounds in activated sludge batch reactors; Stasinakis AS et al.; The biodegradation behavior of four organotin (OT) compounds, namely tributyltin (TBT), dibutyltin (DBT), monobutyltin (MBT) and triphenyltin (TPhT), was studied in lab-scale activated sludge batch reactors. The activated sludge was spiked with the OT compounds at a level of 100mugl(-1) as Sn. Determination of the OT compounds by GC-FPD after ethylation in the dissolved and particulate phase revealed that 24h after the start of the experiments, almost the total of OT compounds has been removed from the dissolved phase and is associated with the suspended solids. Calculation of mass balance in batch reactors showed that OT compounds biodegradation was performed via a sequential dealkylation process. Removals due to biodegradation were differentiated according to the parent compound. In experiments with non-acclimatized biomass, a percentage of 27.1, 8.3, 73.8 and 51.3 was still present as TBT, DBT, MBT and TPhT, respectively, at the end of the experiment (18th day). Half-lives (t(1/2)) of 10.2 and 5.1 days were calculated for TBT and DBT, respectively, whereas apparent t(1/2) values could not be determined for MBT and TPhT (t(1/2)>18 days). The capacity of activated sludge to biodegrade OT compounds in the absence of supplemental substrate indicated that these compounds can be metabolized as single sources of carbon and energy in activated sludge systems. Excluding TBT, the presence of low concentrations of supplemental substrate did not affect the biodegradation potential of activated sludge. The acclimatization of biomass on OT compounds enhanced significantly biodegradation, resulting in significant decreases of half-lives of OT compounds. As a result in the presence of acclimatized biomass, half-lives of 1.4, 3.6, 9.8 and 5.0 days were calculated for TBT, DBT, MBT and TPhT, respectively.

Genes Involved in Aniline Degradation by Delftia acidovorans Strain 7N and Its Distribution in the Natural Environment; Urata M et al.; Aniline-degraders were isolated from activated sludge and environmental samples and classified into eight phylogenetic groups. Seven groups were classified into Gram-negative bacteria, such as Acidovorax sp., Acinetobacter sp., Delftia sp., Comamonas sp., and Pseudomonas sp., suggesting the possible dominance of Gram-negative aniline-degraders in the environment. Aniline degradative genes were cloned from D. acidovorans strain 7N, and the nucleotide sequence of the 8,039-bp fragment containing eight open reading frames was determined. Their deduced amino acid sequences showed homologies to glutamine synthetase (GS)-like protein, glutamine amidotransferase (GA)-like protein, large and small subunits of aniline dioxygenase, reductase, LysR-type regulator, small ferredoxin-like protein, and catechol 2,3-dioxygenase, suggesting a high similarity of this gene cluster to those in P. putida strain UCC22 and Acinetobacter sp. strain YAA. Polymerase chain reaction (PCR) and sequencing analyses of GS-like protein gene segments of other Gram-negative bacteria suggested that Gram-negative bacteria have aniline degradative gene that can be divided into two distinctive groups.

Leucobacter chromiireducens sp. nov, and Leucobacter aridicollis sp. nov., two new species isolated from a chromium contaminated environment; Morais PV et al.; Two strains designated strains L-1T and L-9T were isolated from activated sludge of a treatment plant that receives wastewater from the tannery industry contaminated with chromium. Phylogenetic analysis showed that the organisms represented two new species of the genus Leucobacter. Strains L-1T and L-9T could be distinguished from the type strain of L. komagatae and from the type strain of "L. albus" by the B-type peptidoglycan composition, fatty acid composition, several phenotypic and physiological characteristics. The major fatty acids of the organisms were iso- and anteiso-branched C15:0 and C17:0, straight-chain C16:0 was also found in relatively high proportions. The organisms were halotolerant, grew in medium containing 9% NaCl, and all strains, including the type strain of L. komagatae grew in medium containing 5 mM Cr(VI). On the basis of the distinct peptidoglycan composition, 16S ribosomal DNA sequence analysis, percentage of DNA-DNA reassociation values, and phenotypic characteristics we are of the opinion that strain L-1T represents a new species of the genus Leucobacter for which we propose the name Leucobacter chromiireducens and that strain L-9T represents an additional new species of the same genus for which we propose the name Leucobacter aridicollis.

The solids retention time-a suitable design parameter to evaluate the capacity of wastewater treatment plants to remove micropollutants; Clara M et al.; Micropollutants as endocrine disrupting compounds (EDC) or pharmaceuticals are of increased interest in water pollution control. Wastewater treatment plant (WWTP) effluents are relevant point sources for residues of these compounds in the aquatic environment. The solids retention time (SRT) is one important parameter for the design of WWTPs, relating to growth rate of microorganisms and to effluent concentrations. If a specific substance is degraded in dependency on the SRT, a critical value for the sludge age can be determined. In WWTPs operating SRTs below this critical value, effluent concentrations in the range of influent concentrations or a distribution according to the adsorption equilibrium have to be expected, whereas in WWTPs operating at SRTs higher than the critical value degradation will occur. Critical SRTs were determined for different micropollutants, indicating that the design criteria based on the sludge age allows an estimation of emissions. Different treatment technologies as conventional activated sludge systems and a membrane bioreactor were considered and no significant differences in the treatment efficiency were detected when operated at comparable SRT. The results of the investigations lead to the conclusion that low effluent concentrations can be achieved in WWTPs operating SRTs higher than 10 days (referred to a temperature of 10 degrees C). This corresponds to the requirements for WWTPs situated in sensitive areas according to the urban wastewater directive of the European Community (91/271/EEC) in moderate climatic zones.

Reduction of bromate to bromide coupled to acetate oxidation by anaerobic mixed microbial cultures; van Ginkel CG et al.; Bromate, a weakly mutagenic oxidizing agent, exists in surface waters. The biodegradation of bromate was investigated by assessing the ability of mixed cultures of micro-organisms for utilization of bromate as electron acceptor and acetate as electron donor. Reduction of bromate was only observed at relatively low concentrations (<3.0mM) in the absence of molecular oxygen. Under these conditions bromate was reduced stoichiometrically to bromide. Unadapted sludge from an activated sludge treatment plant and a digester reduced bromate without lag period at a constant rate. Using an enrichment culture adapted to bromate, it was demonstrated that bromate was a terminal electron acceptor for anaerobic growth. Approximately 50%of the acetate was utilized for growth with bromate by the enrichment culture. A doubling of 20h was estimated from a logarithmic growth curve. Other electron acceptors, like perchlorate, chlorate and nitrate, were not reduced or at negligible rates by bromate-utilizing microorganisms.

Fate of Ah receptor agonists during biological treatment of an industrial sludge containing explosives and pharmaceutical residues; Gustavsson LK et al.; GOAL, SCOPE AND BACKGROUND: Sweden is meeting prohibition for deposition of organic waste from 2005. Since 1 million tons of sludge is produced every year in Sweden and the capacity for incineration does not fill the demands, other methods of sludge management have to be introduced to a higher degree. Two biological treatment alternatives are anaerobic digestion and composting. Different oxygen concentrations result in different microbial degradation pathways and, consequently, in a different quality of the digestion or composting residue, It is therefore necessary to study sludge treatment during different oxygen regimes in order to follow both degradation of compounds and change in toxicity. In this study, an industrial sludge containing explosives and pharmaceutical residues was treated with anaerobic digestion or composting, and the change in toxicity was studied. Nitroaromatic compounds, which are the main ingredients of both pharmaceutical and explosives, are well known to cause cytotoxicity and genotoxicity. However, little data are available concerning sludge with nitroaromatics and any associated dioxin-like activity. Therefore, we studied the sludge before and after the treatments in order to detect any changes in levels of Ah receptor (AhR) agonists using two bioassays for dioxin-like compounds. METHODS: An industrial sludge was treated with anaerobic digestion or composting in small reactors in a semi-continuous manner.

The same volume as the feeding volume was taken out daily and stored at -20 degrees C. Sample preparation for the bioassays was done by extraction using organic solvents, followed by clean up with silica gel or sulphuric acid, yielding two fractions. The fractions were dissolved in DMSO and tested in the bioassays. The dioxin-like activity was measured using the DR-CALUX assay with transfected H4IIE rat hepatoma pGudluc cells and an EROD induction assay with RTL-W1 rainbow trout liver cells. RESULTS AND DISCUSSION: The bioassays showed that the sludge contained AhR agonists at levels of TCDD equivalents (TEQs) higher than other sludge types in Sweden. In addition, the TEQ values for the acid resistant fractions increased considerably after anaerobic digestion, resulting in an apparent formation of acid resistant TEQs in the anaerobic reactors.

Similar results have been reported from studies of fermented household waste. There was a large difference in effects between the two bioassays, with higher TEQ levels in the RTL-W1 EROD assay than in the DR-CALUX assay. This is possibly due to a more rapid metabolism in rat hepatocytes than in trout hepatocytes or to differences in sensitivities for the AhR agonists in the sludge. It was also demonstrated by GC/FID analysis that the sludge contained high concentrations of nitroaromatics. It is suggested that nitroaromatic metabolites, such as aromatic amines and nitroanilines, are possible candidates for the observed bioassay effects. It was also found that the AhR agonists in the sludge samples were volatile.

The sludge contained fairly high concentrations of volatile AhR agonists. The increase of acid resistant AhR agonist after anaerobic digestion warrants further investigations of the chemical and toxic properties of these compounds and of the mechanisms behind this observation. RECOMMENDATION AND OUTLOOK: This study has pointed out the benefits of using different types of mechanism-specific bioassays when evaluating the change in toxicity by sludge treatment, in which measurement of dioxin-like activity can be a valuable tool.

In order to study the recalcitrant properties of the compounds in the sludge using the DR-CALUX assay, the exposure time can be varied between 6 and 24 hours. The properties of the acid-resistant AhR agonists formed in the anaerobic treatment have to be investigated in order to choose the most appropriate method for sludge management.

Reduction of excess sludge production by 3,3',4', 5-tetrachlorosalicylanilide in an activated sludge process; Ye FX et al.; The potential of 3,3',4',5-tetrachlorosalicylanilide (TCS) addition to an activated sludge continuous process to reduce excess sludge production by disrupting coupling between anabolism and catabolism was investigated. TCS was chosen as a metabolic uncoupler for continuous test in a lab-scale completely mixed activated sludge process. TCS reduced sludge yield by approximately 30% at a dosage of 40 mg/day. Substrate removal capability was not adversely affected by the presence of TCS, but effluent nitrogen concentration increased during the 60-day continuous operation. Sludge settleability of treated and control samples was qualitatively comparable and not significantly different. Microbial activities in terms of specific oxygen uptake rate were also enhanced, and the microbial population was altered. The results suggest that TCS is an effective chemical uncoupler that reduces sludge yield; process performance was not significantly affected by introduction of the uncoupler.

Stability of the fluorogenic enzyme substrates and pH optima of enzyme activities in different Finnish soils; Niemi RM et al.; Fluorogenic artificial substrates facilitate sensitive enzyme activity measurements for a variety of processes in soil and other environmental samples. It is possible to use in situ pH for measurements on condition that the substrates are chemically stable. We studied the stability of 12 different methyl umbellipherone (MUF) and amino methyl coumarine (AMC) derivatives used as substrates for arylsulphatase, alpha-glucosidase, beta-glucosidase, beta-xylosidase, cellobiosidase, chitinase, phosphomonoesterase (PME), phoshodiesterase (PDE), esterase, lipase and alanine- and leucine aminopeptidases (AP) over the pH range from 4.0 to 8.0 in modified universal buffer (MUB). Stability of the substrates for lipase (4-MUFheptanoate) and esterase (4-MUF-acetate) measurements was poor, especially at the higher pH values. Chitinase substrate, 4-MUF-N-acetyl-beta-d-glucosamide, was unstable at high pH values whereas the substrate for PME activity measurement (4-MUF-phosphate) disintegrated at low pH. The other substrates and MUF and AMC standard solutions were stable over the pH range studied. The optima between pH 4 and 8 of the 11 different enzyme activities were measured in three forest and two agricultural soil samples and in one activated sludge sample. In soil, for alanine and leucine AP the pH optima were usually 7.5 or higher, for arylsulphatase, beta-glucosidase, beta-xylosidase, esterase and PDE between 4 and 5.5, and for cellobiosidase between 4 and 5. alpha-Glucosidase had an optimum below 5.5 but also exhibited high activity at pH 7. Soil-dependent variation in pH optima were observed for chitinase, esterase, PDE and PME. Enzyme activities were also measured in 0.5 M acetate buffer at pH 5.5. This buffer yielded the highest activities in all soil samples for arylsulphatase, PDE and PME.

Persistent organic pollutants (POPs) in the conventional activated sludge treatment process: fate and mass balance; Katsoyiannis A et al.; The fate and the mass balance of persistent organic pollutants (POPs) during the conventional activated sludge treatment process were investigated in the wastewater treatment plant of the city of Thessaloniki, northern Greece. The POPs of interest were 7 polychlorinated biphenyls and 19 organochlorine pesticides. Target compounds were determined at six different points across the treatment system: the influent, the effluent of the primary sedimentation tank, the effluent of the secondary sedimentation tank, the primary sludge, the activated sludge from the recirculation stream, and the digested/dewatered sludge. The distribution of POPs between the dissolved and the adsorbed phases of wastewater and sludge was investigated. A good linear relationship between the distribution coefficients, K(d), and the octanol-water partition coefficients, K(ow), of the solutes was observed only in raw wastewater, suggesting that other factors affect the phase distribution of organic compounds in treated wastewater. For all POPs, a significant increase in partitioning with a decreasing solids concentration was observed, revealing an effect from non-settling microparticles remaining in the "dissolved" phase during the separation procedure. A good linear relationship was also revealed between logK(d) and the dissolved organic carbon (DOC) content of wastewater, suggesting that DOC favors the advective transport of POPs in the dissolved phase. Almost all POPs showed good mass balance agreements at both the primary and the secondary treatment. The losses observed for some species could be attributed to biodegradation/biotransformation rather than volatilization. The relative distribution between the treated effluent and the waste sludge streams varied largely among different compounds, with p-p′-DDE being highly accumulated in the waste sludge (98%) but almost 60% of alpha-HCH remaining in the treated effluent.

Recovery of energy from sludge—comparison of the various options; Rulkens WH et al.; This paper gives a general discussion of existing sludge treatment processes and of innovative treatment options, especially focused on the valuable use of organic carbon compounds as an energy source. Attention is paid both to high and to low temperature processes and to combinations of these processes. Based on an assessment and comparison the most beneficial and sustainable options are identified and a rough indication is given of the possibilities for improvement and future developments.

Sludge as source of energy and revenue; Onyeche TI; Sludge is a residue/product from wastewater treatment plants and contains most of the contaminants released during human activities. Some stringent environmental regulations on sludge treatment and disposal exist in many countries. This has resulted in increasing interest in sludge treatment methods that encourage sludge reduction and improvement in biogas production during anaerobic digestion processes. This work demonstrates the first exploitation of valuable energy from homogenised sludge at technical scale with mass reduction. The optimal combination of sludge homogenisation at relatively low pressures using a modified high-pressure homogeniser led to the success of this unique project. Results showed that about 30% more energy could be obtained from thickened and disrupted sludge than from untreated samples. The energy produced was higher than that invested during disruption and digestion processes. About 23% sludge reduction was also observed with no increase in chemical oxygen demand. This new process can produce extra energy for local electrification and heating the digester while the sludge reduction provides economic benefits. Concentration of sludge causes reduction in investment cost on digester as well as reduction in operational time for sludge dewatering.

Direct generation of electricity from sludges and other liquid wastes; Dentel SK et al.; Recently it has been shown that electrical energy can be harvested from marine sediments, simply by connection of an electrode (anode) in anaerobic marine sediments to an electrode in the aerobic zone above the sediments. We have now shown that similar applications are available in sludge treatment.

Using a reactor with graphite foil electrodes in an aerated aerobic and anaerobic sludge zone, electrical current was generated, and enhanced when an additional organic substrate (acetate) was added. Electron microscopy, x-ray diffraction, and PCR examination of the anode surface showed no surface colonization and no increase in Geobacter relative to a control, indicating that microbial use of the anode as an electron acceptor wa indirect through the use of redox mediators. Given the demonstration of electricity generation from sludge, the potential for similar applications, using other organic waste sources, is evaluated.

Physical characteristics of a waste activated sludge: conditioning responses and correlations with a synthetic surrogate; Dursun D et al.; The efficient and economic management of waste activated sludge (WAS) requires a proper understanding of the sludge's material properties. Though there has been much study of WAS, an adequate linkage between its physical and chemical properties has been elusive. In particular, the conditioning and dewatering of WAS are expensive operations, and the addition of polymer to WAS leads to even more complex material behavior that is difficult to optimize. This paper reports on an extensive characterization of WAS, both with and without polymer conditioning. We combined the classical "jar test" approach with less conventional rheometric and electrokinetic measurements. In addition to the use of sampled WAS, a synthetic surrogate sludge was formulated, attempting to duplicate properties of WAS and allow more extensive characterization of a reproducible surrogate to WAS. Results with both the synthetic surrogate and WAS indicated that the traditional, electrokinetic, and rheological properties were related. However, the dose optima by rheometry were somewhat higher in both cases.

Minimization of sludge production and stable operational condition of a submerged membrane activated sludge process; Bhatta CP et al.; Modification of conventional activated sludge process by adding microfiltration type hollow fiber membrane module offers various advantages. The membrane module is immersed directly in the bioreactor to separate the treated water from the sludge and the modified process is called submerged membrane activated sludge process (SMASP). Bench scale SMASP units were operated at constant flux to study the possibility of minimizing the excess sludge production and stable operation at different SRT and BOD loading. The long SRT developed high stabilized MLSS concentration in the bioreactor and decreased the sludge loading rate {kg-BOD/kg-MLSS-d}. In SMASP, very low sludge yield coefficients (0.04-

0.09 kg-MLSS/kg-BOD) were obtained at long SRT (500 d) and small BOD loading (< or =0.5 kg/(m3 x d)) as compared to conventional activated sludge process and these coefficients showed considerable minimization in excess activated sludge production. At the same operation conditions, suction pressure was also maintained low (<25 kPa), and then SMASP was operated stably for long time (approximately 200 d).

Co-conditioning and dewatering of alum sludge and waste activated sludge; Lai JY et al.; Co-conditioning and dewatering behaviors of alum sludge and waste activated sludge were investigated. Two different sludges were mixed at various ratios (2:1; 1:1; 1:2; 1:4) for study. Capillary suction time (CST) and specific resistance to filtration (SRF) were utilized to assess sludge dewaterability. Relatively speaking, waste activated sludge, though of higher solid content, was more difficult to be dewatered than alum sludge. It was found that sludge dewaterability and settlability became better with increasing fraction of alum sludge in the mixed sludge. Dosage required of the cationic polyelectrolyte (KP-201C) for dewatering was reduced as well. It is proposed that alum sludge acts as skeleton builder in the mixed sludge, and renders the mixed sludge more incompressible which is beneficial for sludge dewatering. e, a, g. Implications of the results of the study to the sludge management plan for Taipei City that generates both alum sludge and waste activated sludge at significant amount are also discussed. The current sludge treatment and disposal plan in Metropolitan Taipei could be made more cost-effective.

Ultrasonic sludge treatment for enhanced anaerobic digestion; Hogan F et al.; Ultrasound is the term used to describe sound energy at frequencies above 20 kHz. High-powered ultrasound can be applied to a waste stream via purpose-designed tools in order to induce cavitation. This effect results in the rupture of cellular material and reduction of particle size in the waste stream, making the cells more amenable to downstream processing. sonix is a new technology utilising high-powered, concentrated ultrasound for conditioning sludges prior to further treatment. This paper presents recent results from a number of demonstration and full-scale plants treating thickened waste activated sludge (TWAS) prior to anaerobic digestion, therefore enhancing the process. The present studies have proved that the use of ultrasound to enhance anaerobic digestion can be achieved at full scale and effectively result in the TWAS (typically difficult to digest) behaving, after sonication, as if it were a "primary" sludge. The technology presents benefits in terms of increased biogas production, better solids reduction, improved dewatering characteristics of the digested sludge mixture and relatively short payback periods of two years or less subject to the site conditions and practices applicable at that time.

Three-dimensional dual-morphotype species modeling of activated sludge flocs; Martins AM et al.; An individual-based model, originally developed for a biofilm system, was adapted to simulate three-dimensional formation of activated sludge flocs. The model was extended to two different bacterial morphotypes (floc-forming and filamentous bacteria), allowing spatial development of the floc according to the bacterial morphology, diffusion, reaction, and growth processes. The model needed also extension with a process for attachment of individual cells. Despite being in an early stage of development, the model is already a tool that enables us to obtain useful information about the microfloc environment. The model indicates that filamentous bacterial morphology and substrate microgradients are important aspects in the formation of bacterial structures. In mass transport-limited regimes filamentous bacterial structures prevail, whereas in growth-limited regimes irregular shaped flocs with fingerlike structures are dominant. These modeling results suggest that activated sludge flocs and biofilms might be different manifestations of the same phenomena. The model results support the hypothesis that floc-macrogradients can be the most important parameter for development of bulking sludge. The model suggests that attachment has a very strong effect on floc structure, leading to enhancement of the effect of substrate microgradients.

Physicochemical properties and stability of activated sludge flocs under temperature upshifts from 30 to 45 degrees C; Morgan-Sagastume F et al.; The impacts of temperature shifts from 30 to 45 degrees C on the structural stability and surface charge of activated sludge flocs were assessed in four sequencing batch reactors (SBRs) treating pulp and paper mill effluent. The improvement in floc stability was tested by sludge magnesium enrichment in one SBR and by operating another reactor at a high sludge retention time (SRT) of 33 days. Floc stability was characterized by dissociation constants with solutions of CaCl(2), KCl, urea, and ethylenediamine tetraacetate (EDTA). Surface charge was assessed by cationic-anionic titration and metals concentrations were also determined. The temperature shift consistently caused an increase in the negative sludge surface charge from approximately -0.180 to -0.300 meq/g MLSS. Magnesium enrichment and a high SRT of 33 days promoted less negatively charged sludge, dampened the increase in negative sludge surface charge, and yielded structurally stronger flocs; however, sludge deflocculation still occurred. Manganese and iron appeared to be released by sludge under the temperature shift. It was concluded that the temperature shift deteriorates the flocculating physicochemical properties of the sludge and that better floc stability achieved by magnesium enrichment and a high SRT is not enough to stop deflocculation. Further research is required to clarify the origin of the increase in negative sludge surface charge, the role of metals, and the governing factors in sludge deflocculation under such temperature shifts.

Optimized sequence retrieval from single bands of temperature gradient gel electrophoresis profiles of the amplified 16S rDNA fragments from an activated sludge system; Zhang X et al.; Sequence retrieval from single bands of polymerase chain reaction (PCR)-denaturing gel electrophoresis (DGE) profiles is an important but often difficult step for molecular diversity analysis of complex microbial communities such as activated sludge systems. We analyzed the temperature gradient gel electrophoresis (TGGE) profiles of PCR-amplified 16S rDNA fragments from an activated sludge sample of a coking wastewater treatment plant. Single bands were excised, and a clone library was constructed for each. Sequence heterogeneity in each single band was found to be significantly overestimated due to single-stranded DNA (ssDNA) contamination formed during the PCR amplification, since only 10-60% of library clones of each single TGGE band had identical migration behavior compared with the parent band.

Three methods, digestion with mung bean nuclease, optimization of PCR amplification, and purification via denatured polyacrylamide gel electrophoresis (d-PAGE), were compared for their ability to minimize ssDNA contamination, with the last one being the most efficient. After using d-PAGE to minimize ssDNA to a nearly nondetectable level, 70-100% of library clones for each single TGGE band had identical migration compared with the parent band. Several sequences were found in each of six single bands, and this co-migration could be predicted with the Poland software. The predominant bacteria of the activated sludge were assessed via a combination of sequence retrieval from each single TGGE band and band intensity analysis. Only beta and alpha subclasses of the Proteobacteria were detected, 93.8% and 6.2%, respectively. Our work suggests that prior to constructing a clone library to retrieve the actual sequence diversity of a single DGE band, it is advisable to minimize ssDNA contamination to a nondetectable level.

High rate treatment by aerobic upflow sludge blanket (AUSB) with external oxygenation; Sharma KR et al.; A 3-year study was conducted using an aerobic upflow sludge blanket (AUSB) reactor to achieve high-rate biological treatment through maintenance of a high biomass concentration (7-10 g VSS/L) together with a high oxygen flux.

The AUSB reactor was not aerated directly; instead, oxygenation was provided in an external chamber with flow recirculation. The oxygenation was provided at four different pressurizations: 0, 15, 20 and 25 psig.

The AUSB reactor was also pressurized to avoid the flotation of biomass. The flow recirculation rate was varied from 400% to 1,500% in order to ensure adequate oxygen supply. It was found that the AUSB system was able to handle a volumetric loading of as high as 10 kg COD/m3-day with a removal efficiency of 92%. Despite a high upflow velocity through AUSB, the effluent suspended solid concentration was mostly below 60 mg/L.

The active fraction of biomass in the AUSB sludge was about 3 times higher than that of the regular activated sludge. This was indicated by a very high specific oxygen uptake rate (SOUR), up to 180-250 mg O2/g VSS-hr. The sludge yield in the entire system was only 0.09 to 0.13 g VSS/g COD removed. This was mainly caused by additional auto-oxidation of biosolids in the oxygenation chamber due to flow recirculation.

Effect of chlorination bulking control on water quality and phosphate release/uptake in an anaerobic-oxic activated sludge system; Chang WC et al.; This study evaluates the effect of chlorination bulking control on water quality and phosphate release/uptake in an anaerobic-oxic activated sludge system. A series of batch experiments with different specific NaOCl mass dose were conducted to determine the sludge settling properties, supernatant water quality and phosphate metabolism behavior of filamentous bulking sludge.

The harvested sludge was from a continuous-flow anaerobic-oxic (A/O) activated sludge pilot-plant, i.e., enhanced biological phosphorus removal (EBPR) system, operated with

15 days of sludge retention time. The filamentous bacteria in the A/O pilot plant were identified to be Thiothrix according to Eikelboom's classification techniques, which was in accordance with the high influent sulfate concentration of this study (50 mg/L sulfate).

Increasing NaOCI concentration, as revealed by experimental results, obviously decreased the sludge settling properties (SVI values and zone settling velocities) and meanwhile significantly reduced supernatant water quality (COD, SS, TP) mainly due to higher suspended solids caused by floc disruption.

Moreover, the nine-hour batch experiments indicated that high NaOCI dosage (40 mg/gMLSS) completely deteriorated phosphate metabolism of EBPR sludge. Such a high dosage of chlorination further confirmed overdosing through disappearance of intracellular PHB and death of protozoa by microscopic investigation.

Still, phosphate release/uptake behavior of EBPR sludge properly functions at low NaOCl dosage (5 mg/g MLSS). Besides, phosphate metabolism worsens rapidly before the SVI value reaches its lowest level. These findings imply that determining NaOCI requirement with merely SVI values can readily result in chlorination overdosing. Proper NaOCI dosage requires a delicately balanced consideration between sludge settling improvement, water quality demand and phosphate metabolism. Batch test of phosphate release/uptake is apparently a prerequisite to conclude an appropriate NaOCl dosage for bulking control.

Microbial degradation of estrogens using activated sludge and night soil-composting microorganisms; Shi JH et al.; In order to investigate the potential for microbial degradation of estrogens, and the products formed, activated sludge collected from Korea (ASK) and night soil-composting microorganisms (NSCM) were used to degrade estrogens. Results showed that both ASK and NSCM degraded almost 100% of the natural estrogens estrone (E1), 17beta-estradiol (E2), and estriol (E3) from initial concentrations of 20-25 mg/L, while synthetic estrogen, ethynylestradiol (EE2), was not degraded.

Analysis of degradation products of E2 by using HPLC-ECD and a consecutive first-order reaction calculation confirmed that E2 was sequentially degraded to E1, which was further degraded to other unknown compounds by ASK and NSCM.

We then used the yeast two-hybrid assay to show that the unknown degradation products did not appear to possess estrogenic activity when E1, E2 or E3 were degraded to below the detection limit after 14 days of incubation, indicating that ASK and NSCM not only degrade natural estrogens, but also remove their estrogenic activities.

Effects of phosphorus limitation and temperature on PHA production in activated sludge; Chinwetkitvanich S et al.; The study was designed to investigate the effects of temperature and phosphorus limitation on polyhydroxyalkanoate (PHA) production and storage by activated sludge biomass. The two-stage operation approach, i.e. a growth phase followed by a nutrient limitation phase, was applied to induce PHA accumulation.

The pre-selected temperatures of 10, 20 and 30 degrees C were investigated under phosphorus limitation conditions using three four-litre fully aerobic SBR systems operated at an SRT of 10 days with cycle time and HRT of 6 and 10 hours.

PHA production was greater in the 10 degrees C system than in the 20 degrees C and 30 degrees C systems but there was little difference between the two higher temperatures. The maximum PHA fractions of the sludge were 52, 45 and 47%TSS for the three temperatures from low to high, and the maximum PHA concentrations in the mixed liquors were 1,491, 1,294 and 1,260 mg/l, respectively. However, it was observed that very low values of PHA yield per unit COD consumed were obtained, i.e., 0.05, 0.03 and 0.04 mgPHA/mgCODu, for the 10, 20 and 30 degrees C reactors, respectively.

This was because all three systems required several days to reach maximum PHA accumulation in their mixed liquor biomasses. It is probable the bacteria still had some stored poly-P in their cells upon initiation of the phosphorus limited influent, and PHA accumulation was delayed until the stored phosphorus was depleted. Also, PHA productivity was reduced by the large amounts of biomass lost from the systems because of sludge bulking.

Sludge thickening performance of mesh filtration process; Park MS et al.; Small-scale wastewater treatment facilities play an important role in improving the aquatic environment in many countries.

Although sludge treatment is essential for overall wastewater treatment, it is difficult for small-scale facilities to use mechanical equipment or other facilities.

As the first step of the sludge treatment, it is important to develop a convenient sludge thickening process for small-scale facilities. In this work, we examined the sludge thickening performance of a mesh filtration system: the mesh opening sizes of 100-500 microm, and the sludge (3,000-9,000 mg-SS/L) was obtained from a domestic wastewater treatment facility. The filtration was carried out only under the hydraulic pressure between the water level and the effluent port connected to the mesh filter module.

The sludge reduction rates were in the range of 85-95% for 6-7 h; the initial filtration rate was very high, but the rate decreased with a decrease in hydraulic pressure due to the reduction of the water level in the vessel. In addition, the effluents (passed through the mesh) contained very low SS and could be directly discharged into the environment.

Biological treatability of raw and ozonated penicillin formulation effluent; Cokgor EU et al.; In the present study, oxidative pre-treatment of pharmaceutical wastewater originating from the formulation of the penicillin Sultamycillin Tosylate Diydrate via ozonation at varying pH and ozone feed rates was investigated. Biological treatability studies were performed with a synthetic wastewater alone and supplemented with raw and ozonated penicillin formulation effluents.

The highest COD (34%) and TOC (24%) removal efficiencies were obtained at pH 11.0, whereas the BOD5 value increased from 16 mg l(-1) to 128 mg l(-1) after 40 min of ozonation, corresponding to an applied ozone dose of 1670 mg l(-1) and 33% relative ozone absorption. The studies showed that no degradation of raw penicillin fraction (30% of total COD) occurred, and degradation of the synthetic wastewater being completely treatable without penicillin addition, was inhibited by 7%.

Upon 40 min ozonation, the synthetic wastewater could be completely oxidized and at the same time 35% of ozonated penicillin wastewater removal was obtained. Respirometric studies were conducted in parallel and produced results indicating a 22% decrease in the total oxygen consumption rate established for raw penicillin formulation effluent compared to the results obtained from the aerobic batch reactor. No inhibition of the synthetic fraction was observed for the 40 min-ozonated penicillin formulation effluent, biodegradability of the 60 min-ozonated penicillin effluent decreased possibly due to recalcitrant oxidation product accumulation. The modeling study provided experimental support and information on inhibition kinetics in activated sludge model no. 3 (ASM3) by means of respirometric tests for the first time.

Reduction of coagulant amount added to activated sludge for phosphorus removal; Nakajima J et al.; Adding coagulant to the activated sludge process is effective in maintaining the stability of phosphorus removal. However, the precise mechanisms of the reaction and behavior of coagulants and phosphorus are not well known.

By introducing a new phosphorus removal model (PRM), the behavior of coagulant and phosphorus in the process could be described. The experimental data of the effluent phosphorus concentration and Fe content in the activated sludge agreed with the values calculated by PRM. The amount of coagulant addition to the activated sludge process for phosphorus removal is reduced with the enhanced biological phosphorus removal process. It is suggested that the amount of reduction is determined by using PRM.

Monitoring activated sludge settling properties using image analysis; Jenne R et al.; The goal of this study is to develop a monitoring system for activated sludge properties, as this is an essential tool in the battle against filamentous bulking. A fully automatic image analysis procedure for recognising and characterising flocs and filaments in activated sludge images has been optimised and subsequently used to monitor activated sludge properties in a lab-scale installation.

The results of two experiments indicate that the image information correlates well with the Sludge Volume Index. It is shown that, at the onset of filamentous bulking, there is an increase in total filament length on the one hand, and a change in floc shape on the other hand.

Performances of three R-AN-D-N wastewater treatment plants in the Czech Republic; Bejvl Z et al.; The evolution of Czech standards requires higher efficiency of nutrient removal from municipal wastewaters. At the beginning of the last decade of 20th century, a new activated sludge configuration called R-AN-D-N process has been described, successfully tested and now largely used at several wastewater treatment plants (WWTP) in the Czech republic. The main feature of the R-AN-D-N process is the introduction of a regeneration zone in sludge recycle, which enables to increase sludge age in the system without any substantial increase in WWTP volume. Performances of three Czech large WWTP with R-AN-D-N configuration have been monitored and compared within a period of one and a half years.

The results confirmed excellent nutrient removal efficiency for wastewaters with different proportion between sewage and industrial effluents. Two of three monitored WWTP received wastewaters from breweries (Budweiser and Pilsner Urquell). The settleability of activated sludge from all three WWTP was correct, with SVI values usually ranging from 50 to 150 ml/g. Monitoring of sludge composition indicated proliferation of several filamentous bacteria, particularly types 0581, 0092 and M. parvicella. No severe bulking events were observed. Finally, the operational costs expressed in CZK (Czech crown: 1 CZK = {see symbol in text}0.0322) per cubic metre of treated sewage or per capita amounts respectively from 2.24 to 6.52, and from 285 to 342.

2

Environmental Biotechnology in New Era

North American performance experience with anoxic and anaerobic selectors for activated sludge bulking control; Parker D et al.; Obtaining reliably high compaction characteristics and low SVI values has enormous economic consequences on activated sludge plant design and operation. Now that use of selectors has become more common for achieving low SVI values, the assessment of the full-scale performance characteristics of activated sludge plants incorporating selectors has high value for the design and operation of future plants.

Treatment plant effluent quality is seldom analyzed solely on an average basis and the impact peak events have on effluent quality must be considered in design and operation. Descriptive statistics should also be used to assess selector performance, so that the influence of infrequently occurring values on plant design and operation can be assessed. The performance of 21 activated sludge plants incorporating selectors was evaluated. All the selector plants for which post installation data were available showed that the operating SVI values were significantly improved. One plant studied in detail showed that final effluent quality was significantly better after the selector installation. As a group, activated sludge plants with anaerobic selectors outperform those with anoxic selectors. Dissolved oxygen control is just as important in activated sludge plants with selectors as in conventional activated sludge plants.

Influence of sludge properties and hydraulic loading on the performance of secondary settling tanks—full-scale operational results; Vestner RJ et al.; Full-scale investigations at a WWTP with a two-stage secondary settling tank process revealed relationships between significant operating parameters and performance in terms of effluent suspended solids concentration. Besides common parameters (e.g. surface overflow rate and sludge volume loading rate) feed SS concentration and flocculation time must be considered. Concentration of the return activated sludge may help to estimate the performance of existing secondary settling tanks.

Modeling of a large-scale wastewater treatment plant for efficient operation; Gokcay CF et al.; Environmental legislations in the Western world impose stringent effluent quality standards for ultimate protection of the environment. This is also observed in Turkey. The current paper presents efforts made to simulate an existing 0.77 million m3/day conventional activated sludge plant located at Ankara, AWTP. The ASM1 model was used for simulation in this study. The model contains numerous stoichiometric and kinetic parameters, some of which need to be determined on case by case bases. The easily degradable COD (S(S)) was determined by two methods, physical-chemical and respirometric methods, namely. The latter method was deemed unreliable and rejected in the further study. Dynamic simulation with SSSP program predicted effluent COD and MLSS values successfully while overestimating OUR. A complete fit could only be obtained by introducing a dimensionless correction factor (etaO2 = 0.58) to the oxygen term in ASM1.

A control strategy for reducing aeration costs during low loading periods; Sahlmann C et al.; The efficiency of the aeration system in a full-scale activated sludge basin with 3 separately controlled aeration zones was improved for the low loading period in summer. The air flow rate to each aeration zone is currently regulated to hold a preset dissolved oxygen concentration (DO). Four different DO setpoint combinations were tested, each one for a one week period, using dynamic off-gas testing to measure the standardised oxygen transfer efficiency (alphaSOTE). As the DO setpoints were lowered, the total air flow rate to the basin decreased initially. A low DO in the first zones slowed biomass activity and pushed the load towards the end of the aeration basin. The relationship between alphaSOTE and the specific diffuser flow rate qD is different for each zone. In Zone 1 there was a strong decrease in alphaSOTE as qD increased, while Zones 2 and 3 were fairly independent of qD, Zone 2 at a higher level than Zone 3. Aeration costs were reduced by 15% for the most efficient combination. To achieve even more savings, a control strategy adjusting oxygen transfer rates over the aeration basin to the necessary oxygen transfer rates is suggested. It is based on changing the DO setpoints to reach the lowest total air flow rate while meeting the effluent requirements.

Nutrient removal process development and full scale implementation at the 4 million p.e. main treatment plant of Vienna, Austria; Kroiss H et al.; The Main Treatment Plant of Vienna is in extension for 4 million p.e. and very stringent nutrient removal requirements. The existing high rate BOD removal activated sludge plant (in operation since 1980) is extended by a second stage activated sludge plant and a newly developed flow scheme for nitrogen removal optimisation adaptable to the temperature variations over the year. For this plant pilot investigations have been performed for the development of a specific mathematical model (ASMV) and a specific aeration control strategy. The civil work of the extension is already finished and the installation of the equipment has started. Operation should start in 2004. The whole project will cost about {see symbol in text}264 million of which about one half is for civil work. The effluent standards correspond to the requirements for sensitive areas in EU Directive for Municipal Waste Water. The raw primary and excess sludge are incinerated after thickening and dewatering. This paper tries to condense the already existing literature with the construction progress and the cost situation.

Isolation and characterization of sulphur-oxidizing Thiomonas sp. and its potential application in biological deodorization; Chen XG et al.; AIMS: To isolate and characterize a sulphur-oxidizing bacterial strain from activated sludge and to evaluate its potential application in biological deodorization. METHODS AND RESULTS: A dominant sulphur-oxidizing bacterial strain, designated as strain SS, was isolated from an enrichment culture using thiosulphate as a sole energy source and CO2 as a sole carbon source. The cells of this organism were aerobic, rod-shaped, Gram-negative and motile. Strain SS could grow autotrophically, heterotrophically as well as mixotrophically. Autotrophic growth was observed at pH values ranging from 2.3 to 9.0. Phylogenetic analyses revealed that strain SS belonged to Group 1 of the genus Thiomonas, closely related to Thiomonas perometabolis and Thiomonas intermedia. The thiosulphate oxidation rates of strain SS at different pH values were evaluated in terms of oxygen uptake using a Micro-Oxymax respirometer. The results showed that the maximum oxidation rate of 5.65 mg l(-1) h(-1) occurred at 56 h of growth and pH 6.0. Continuous H2S removal study demonstrated that strain SS could remove more than 99% of H2S when the inlet concentration was below 58.6 ppm. Further increase of the inlet concentration to 118 ppm gave rise to a decline in the removal efficiency to ca 90%. CONCLUSIONS: The strong acidification of the culture medium during the later period could result in the deterioration of the growth activity and the metabolism activity of strain SS. In practical application, the problems caused by the end-product inhibition and the acidification can be alleviated by periodical replacement of culture medium with fresh medium. d, b. Given the physiological flexibility and the ability to remove H2S rapidly and efficiently, strain SS could be a good 'deodorizing' candidate. SIGNIFICANCE AND IMPACT OF THE STUDY: This is the first time that Thiomonas species has been reported for biological deodorization application.

Start up of deammonification process in one single SBR system; Li X et al.; A process for autotrophic nitrogen removal named aerobic/anoxic deammonification wherein NH4+ is oxidized by nearly 50% to NO2- and subsequently the ammonia is converted together with the nitrite to molecular nitrogen (N2 gas), has come to full-scale application within the last few years. In this research, sludge from a biological rotation disk located at a landfill leachate plant at Mechernich, Germany, which is capable of performing the deammonification process, was used as seed sludge for acclimating deammonification activities in laboratory scale batch-reactors. In parallel, the same tests were performed with normal activated sludge. Research results indicated that deammonification activities could be obtained from the seeded reactor and also, with limited performance, from normal activated sludge in a single SBR system after several months acclimation. It was also seen that oxygen is an important factor that influences the deammonification from both the acclimatization process and process running.

Further results were approved that report an impact of nitrite as a process intermediate on the closely related process of anaerobic ammonia oxidation ("Anammox"). However, limiting concentrations on a bacteria population performing deammonification were found to be different to those reported for a pure Anammox-culture. Also the influence of another intermediate, hydrazine, was tested for speeding up the acclimating process by inducing the deammonification activities and recovering the activities of deammonification from nitrite inhibition.

Results of a technology demonstration project to compare rapid aquatic toxicity screening tests in the analysis of industrial effluents; Daniel M et al.; The results of a 'BioWise' demonstration project to assess the comparative sensitivity and practicality of seven new assays for the direct assessment of ecotoxicity in industrial effluents are presented. In addition the aim of the project was to validate the results of the new assays against benchmark data generated from non-proprietary, rapid, microplate screening assays using the regulatory species; freshwater crustacean Daphnia magna and green algae Selenastrum capricornutum, chosen in view of their environmental relevance. The new commercial test assays were: Daphnia magna, Selenastrum capricornutum and Thamnocephalus platyurus Toxkits supplied by Vickers Laboratories Ltd, containing dormant, immobilised life stages of the test species; GreenScreen EM, a yeast based assay for genotoxicity and general acute toxicity supplied by Gentronix Ltd; and CellSense a mediated, amperometric whole cell biosensor based on immobilised activated sludge and E. coli. 38 effluent samples supplied by members of SOCSA (Specialised Organic Chemicals Sector Association) were examined over a period of 13 months, in the project co-ordinated by the AstraZeneca Brixham Environmental Laboratory, and part funded by BioWise via the UK Government Department of Trade and Industry.

Identification and occurrence of tetrad-forming Alphaproteobacteria in anaerobic-aerobic activated sludge processes; Wong MT et al.; In an acetate-fed anaerobic-aerobic membrane bioreactor, a deteriorated enhanced biological phosphorus removal (EBPR) community was developed (as determined based on the chemical profiles of organic substrate, soluble phosphate, and intracellular carbohydrate and polyhydroxyalkanote (PHA) concentrations). Microscopic observations revealed the dominance of tetrad-forming organisms (TFOs), of which the majority stained positively for

PHA under anaerobic conditions. Fluorescence in situ hybridization (FISH) confirmed that the Alphaproteobacteria (85.0+/-7.0% of total cells) were the most dominant group. A 16S rRNA gene clone library specific for the Alphaproteobacteria indicated that most 16S rRNA gene clones (61% of total clones) were closely affiliated with 'Defluvicoccus vanus', forming a cluster within subgroup 1 of the Alphaproteobacteria. Combined PHA staining and FISH with specific probes designed for the members of the 'Defluvicoccus' cluster suggested diversity within this TFO cluster, and that these TFOs were newly identified glycogen-accumulating organisms in EBPR systems. However, these 'Defluvicoccus'-related TFOs were only seen in low abundance in 12 different EBPR and non-EBPR systems, suggesting that they were not the key populations responsible for the deterioration of full-scale EBPR processes.

Conceptualizing "suicidal genetically engineered microorganisms" for bioremediation applications; Pandey G et al.; Use of genetically modified microorganisms (GEMs) for pollution abatement has been limited because of risks associated with their release in the environment. Recent developments in the area of recombinant DNA technologies have paved the way for conceptualizing "suicidal genetically engineered microorganisms" (S-GEMS) to minimize such anticipated hazards and to achieve efficient and safer bioremediation of contaminated sites. Our strategy of designing a novel S-GEM is based on the knowledge of killer-anti-killer gene(s) that would be susceptible to programmed cell death after detoxification of any given contaminated site(s).

Ochrobactrum tritici strain 5bvl1 — characterization of a Cr(VI)-resistant and Cr(VI)-reducing strain; Branco R et al.; Bacterial strain 5bvl1, isolated from a chromium-contaminated wastewater treatment plant and identified as Ochrobactrum tritici, was resistant to a broad range of antibiotics, to Cr(VI), Ni(II), Co(II), Cd(II), and Zn(II), and was able to grow in the presence of 5% NaCl and within the pH range 4–10. Characterization showed that strain 5bvl1 could be considered a halotolerant and alkalitolerant microorganism resistant to high concentrations of Cr(VI). This strain was able to grow aerobically in up to 10 mmol.L–1 Cr(VI). Cr(VI) resistance was independent of sulphate concentration. Under aerobic conditions strain 5bvl1 was also able to reduce high Cr(VI) concentrations (up to 1.7 mmol.L–1). Increasing concentrations of Cr(VI) in the medium lowered the growth rate of strain 5bv11 but the reduction in growth rate could not be directly correlated with the amount of Cr(VI) reduced. Unlike the type strain, which was only able to reduce Cr(VI), strain 5bvl1 was resistant to Cr(VI) and able to reduce it. Moreover, in strain 5bvl1, the rate and extent of Cr(VI)-reduction were higher than in the other strains of the genus Ochrobactrum. Ochrobactrum strain 5bvl1 resists high Cr(VI) concentrations and has a high Cr(VI)-reducing ability, making it a valuable tool in bioremediation.

Bioremediation of a weathered and a recently oil-contaminated soils from Brazil: a comparison study; Trindade PV et al.; The facility with which hydrocarbons can be removed from soils varies inversely with aging of soil samples as a result of weathering. Weathering refers to the result of biological, chemical and physical processes that can affect the type of hydrocarbons that remain in a soil. These processes enhance the sorption of hydrophobic organic contaminants (HOCs) to the soil matrix, decreasing the rate and extent of biodegradation. Additionally, pollutant compounds in high concentrations can more easily affect the microbial population of a recently contaminated soil than in a weathered one, leading to inhibition of the biodegradation process. The present work aimed at comparing the biodegradation efficiencies obtained in a recently oil-contaminated soil (spiked one) from Brazil and an weathered one, contaminated for four years, after the application of bioaugmentation and biostimulation techniques.

Both soils were contaminated with 5.4% of total petroleum hydrocarbons (TPHs) and the highest biodegradation efficiency (7.4%) was reached for the weathered contaminated soil. It could be concluded that the low biodegradation efficiencies reached for all conditions tested reflect the treatment difficulty of a weathered soil contaminated with a high crude oil concentration. Moreover, both soils (weathered and recently contaminated) submitted to bioaugmentation and biostimulation techniques presented biodegradation efficiencies approximately twice as higher as the ones without the aforementioned treatment (natural attenuation).

Biodegradation of phenanthrene by the indigenous microbial biomass in a zinc amended soil; Wong KW et al.; Abstract k.w. wong, b.a. toh, y.p. ting and j.p. obbard. 2004.Aims: To study the effect of zinc on the biodegradation of phenanthrene by the microbial biomass in soil. Methods and Results: Uncontaminated soil was amended with zinc and phenanthrene as single or co-contaminants, and microbial metabolic activity was measured using an intracellular dehydrogenase enzyme bioassay over 37 days.

Contaminants were amended at optimum, action and double the action level specified in 'The New Dutch List' (Ministry of Housing, Spatial Planning and Environment, the Netherlands, 2000). Microbial activity in soils with zinc or phenanthrene alone indicated the presence of tolerant, albeit inhibited soil micro-organisms.

A zinc concentration at the optimum level of 140 mg kg(-1) in the co-contaminated soil (phenanthrene at 40 mg kg(-1)) resulted in marginal stimulation of the rate of phenanthrene biodegradation. However, Zn(2+) concentrations at the action and double the action level of zinc (720 and 1440 mg kg(-1)) inhibited phenanthrene degradation. Conclusions: Biodegradation of phenanthrene in soils co-contaminated with zinc at concentrations above the action value is impeded.

Significance and Impact of the Study: Bioremediation efforts to remove polycyclic aromatic hydrocarbon in zinc co-contaminated soils are likely to be constrained.

Stable isotope fractionation analysis as a tool to monitor biodegradation in contaminated acquifers; Meckenstock RU et al.; The assessment of biodegradation in contaminated aquifers has become an issue of increasing importance in the recent years. To some extent, this can be related to the acceptance of intrinsic bioremediation or monitored natural attenuation as a means to manage contaminated sites. Among the few existing methods to detect biodegradation in the subsurface, stable isotope fractionation analysis (SIFA) is one of the most promising approaches which is pronounced by the drastically increasing number of applications. This review covers the recent laboratory and field studies assessing biodegradation of contaminants via stable isotope analysis.

Stable isotope enrichment factors have been found that vary from no fractionation for dioxygenase reactions converting aromatic hydrocarbons over moderate fractionation by monooxygenase reactions (epsilon=-3 per thousand) and some anaerobic studies on microbial degradation of aromatic hydrocarbons (epsilon=-1.7 per thousand) to larger fractionations by anaerobic dehalogenation reactions of chlorinated solvents (epsilon=between -5 per thousand and -30 per thousand). The different isotope enrichment factors can be related to the respective biochemical reactions. Based on that knowledge, we discuss under what circumstances SIFA can be used for a qualitative or even a quantitative assessment of biodegradation in the environment. In a steadily increasing number of cases, it was possible to explain biodegradation processes in the field based on isotope enrichment factors obtained from laboratory experiments with pure cultures and measured isotope values from the field. The review will focus on the aerobic and anaerobic degradation of aromatic hydrocarbons and chlorinated solvents as the major contaminants of groundwater. Advances in the instrumental development for stable isotope analysis are only mentioned if it is important for the understanding of the application.

Biotechnology applied to cultural heritage: biorestoration of frescoes using viable bacterial cells and enzymes; Ranalli G et al.; Abstract g. ranalli, g. alfano, c. belli, g. lustrato, m.p. colombini, i. bonaduce, e. zanardini, p. abbruscato, f. cappitelli and c. sorlini.

2004.Aims: To set up and employ, for the biorestoration of cultural heritage (altered frescoes), an advanced and innovative biotechnology method based on the sequential use of whole viable bacterial cells and specific enzymes.

Methods and Results: The bioremediation intervention consisted of the direct application onto an artwork surface of whole bacterial cells of the Pseudomonas stutzeri A29 strain (bioaugmentation), followed by, in a final step, a purified Protease enzyme. The bioremediation was performed on a Spinello Aretino fresco that had become altered by the animal glue residues of past restoration.

For the reader's interest the fresco is the 14th century Conversione di S. Efisio e battaglia (Conversion of S. Efisio and battle), size 3.5 x 7.8 m at the Pisa Camposanto Monumentale, Italy. An assessment was made of the final costs of the biological tests (whole bacterial cells, enzymes) so as to compare them with other intervention techniques.

Conclusions: A successful innovative biological approach to recover valuable frescoes was set up, and the best conditions for treatment efficiency were identified. Furthermore the cost of the biological cleaning using viable bacterial cells and enzymes (P. stutzeri, Protease, Collagenase, 1 : 3 : 10, ratio respectively) was much lower than that of other conventional methods, making this biotechnology not only very interesting but also very competitive. Significance and Impact of the Study:

New biotechnologies with an innovative, soft approach to the 'biocleaning' and 'biorestoration' of cultural heritage are in constant demand, and our results are clear evidence that such an approach has been achieved; the technique could be of significant importance towards developing other goals.

MetaRouter: Bioinformatics for bioremediation; Pazos F et al.; Bioremediation, the exploitation of biological catalysts (mostly microorganisms) for removing pollutants from the environment, requires the integration of huge amounts of data from different sources. We have developed MetaRouter, a system for maintaining heterogeneous information related to bioremediation in a framework that allows its query, administration and mining (application of methods for extracting new knowledge). MetaRouter is an application intended for laboratories working in biodegradation and bioremediation, which need to maintain and consult public and private data, linked internally and with external databases, and to extract new information from it. Among the data-mining features is a program included for locating biodegradative pathways for chemical compounds according to a given set of constraints and requirements. The integration of biodegradation information with the corresponding protein and genome data provides a suitable framework for studying the global properties of the bioremediation network. The system can be accessed and administrated through a web interface. The full-featured system (except administration facilities) is freely available at Additional material: http://www.pdg.cnb.uam.es/biodeg_net/MetaRouter.

Lag period of 14CO2 evolution from dioctyl sulpho{2,3-14C}succinate in relation to adaptation of bacterium, Comamonas terrigena, to dialkyl esters of sulphosuccinate; Godocikova J et al.; Comamonas terrigena, strain N3H, which was isolated from soil polluted with crude oil products, degraded dioctyl sulphosuccinate, a synthetic commercial surfactant. The primary degradation of this compound, the cleavage of ester bonds between octyl groups and sulphosuccinate, lasted significantly shorter time than the subsequent breakdown of the sulphosuccinate moiety of dioctyl sulpho{2,3-(14)C}succinate. (14)CO(2) evolution had a significant shorter lag period with cells in Tris/phosphate medium, without inorganic sulphate and adapted to surfactant, than unadapted cells. The acceleration of the primary degradation by adapted cells also suggest that some enzymes involved in surfactant degradation are inducible. The bacterium may be useful for bioremediation.

Physiological and molecular genetic analyses of vinyl chloride and ethene biodegradation in Nocardioides sp. strain JS614; Mattes TE et al.; Nocardioides sp. strain JS614 utilizes vinyl chloride and ethene as carbon and energy sources. JS614 could be influential in natural attenuation and biogeochemical ethene cycling, and useful for bioremediation, biocatalysis and metabolic engineering, but a fundamental understanding of the physiological and genetic basis of vinyl chloride and ethene assimilation in strain JS614 is required. Alkene monooxygenase (AkMO) activity was demonstrated in whole-cell assays and epoxyalkane:coenzyme M transferase (EaCoMT) activity was detected in JS614 cell-free extracts. Pulsed-field gel electrophoresis revealed a 290-kb plasmid (pNoc614) in JS614. Curing experiments and PCR indicated that pNoc614 encodes vinyl chloride/ethene-degradation genes. JS614 vinyl chloride/ethene catabolic genes and flanking DNA (34.8 kb) were retrieved from a fosmid clone. AkMO and EaCoMT genes were found in a putative operon that included CoA transferase, acyl-CoA synthetase, dehydrogenase, and reductase genes. Adjacent to this gene cluster was a divergently transcribed gene cluster that encoded possible coenzyme M biosynthesis enzymes. a, g, i. Reverse transcription-PCR demonstrated the vinyl chloride- and ethene-inducible nature of several genes. Genes encoding possible plasmid conjugation, integration, and partitioning functions were also discovered on the fosmid clone.

Review of disposal technologies for chromated copper arsenate (CCA) treated wood waste, with detailed analyses of thermochemical conversion processes; Helsen L et al.; Several alternative methods for the disposal of chromated copper arsenate (CCA) treated wood waste have been studied in the literature, and these methods are reviewed and compared in this paper. Alternative disposal methods include: recycling and recovery, chemical extraction, bioremediation, electrodialytic remediation and thermal destruction. Thermochemical conversion processes are evaluated in detail based on experiments with model compounds as well as experimental and modelling work with CCA treated wood. The latter category includes: determination of the percentage of arsenic volatilised during thermal conversion of CCA treated wood, identification of the mechanisms responsible for arsenic release, modelling of high temperature equilibrium chemistry involved when CCA treated wood is burned, overview of options available for arsenic capture, characterisation of ash resulting from (co-)combustion of CCA treated wood, concerns about polychlorinated dibenzo-p-dioxins/furans (PCDD/F) formation. Finally, the most appropriate thermochemical disposal technology is identified on short term (co-incineration) and on long term (low-temperature pyrolysis or high-temperature gasification).

A PCR method for the detection and differentiation of Lentinus edodes and Trametes versicolor in defined-mixed cultures used for wastewater treatment; Garcia-Mena J et al.; A PCR-based method for the quantitative detection of Lentinus edodes and Trametes versicolor, two ligninolytic fungi applied for wastewater treatment and bioremediation, was developed. Genomic DNA was used to optimize a PCR method targeting the conserved copper-binding sequence of laccase genes. The method allowed the quantitative detection and differentiation of these fungi in single and defined-mixed cultures after fractionation of the PCR products by electrophoresis in agarose gels. Amplified products of about 150 bp for L. edodes, and about 200 bp for T. versicolor were purified and cloned. The PCR method showed a linear detection response in the 1.0 mug-1 ng range. The same method was tested with genomic DNA from a third fungus (Phanerochaete chrysosporium), yielding a fragment of about 400 bp. Southern-blot and DNA sequence analysis indicated that a specific PCR product was amplified from each genome, and that these corresponded to sequences of laccase genes. This PCR protocol permits the detection and differentiation of three ligninolytic fungi by amplifying DNA fragments of different sizes using a single pair of primers, without further enzymatic restriction of the PCR products. This method has potential use in the monitoring, evaluation, and improvement of fungal cultures used in wastewater treatment processes.

Hydraulics of recirculating well pairs for ground water remediation; Cunningham JA et al.; Recirculating well pairs are a proven means of implementing bioremediation and may also be useful for applying other in situ ground water remediation technologies. A bromide tracer test was performed to characterize the hydraulic performance of a recirculating well pair installed at Moffett Field, California. In particular, we estimate two important properties of the recirculating well pair: (1) the fraction of captured water that is recycled between the wells, and (2) the travel-time distribution of ground water in the induced zone of recirculation. We also develop theoretical estimates of these two properties and demonstrate they depend upon a dimensionless pumping rate, denoted xi. The bromide breakthrough curve predicted from theory agrees well with that determined experimentally at Moffett Field. The minimum travel time between the wells is denoted t(min). In theory, t(min) depends inversely on Q, the pumping rate in the recirculating wells, and is proportional to a2, the square of the distance between the wells. Both the experimental and theoretical travel-time distributions indicate that at least half the recirculating water travels between the wells along fast flowpaths (travel time < 2*t(min)). Therefore, when designing recirculating well pairs, engineers should ensure that t(min) will be sufficiently high to allow biologically mediated reactions (or other in situ remediation processes) sufficient time to proceed.

The enzymes of oxalate metabolism: unexpected structures and mechanisms; Svedruzic D et al.; Oxalate degrading enzymes have a number of potential applications, including medical diagnosis and treatments for hyperoxaluria and other oxalate-related diseases, the production of transgenic plants for human consumption, and bioremediation of the environment. This review seeks to provide a brief overview of current knowledge regarding the major classes of enzymes and related proteins that are employed in plants, fungi, and bacteria to convert oxalate into CO(2) and/or formate. Not only do these enzymes employ intriguing chemical strategies for cleaving the chemically unreactive C-C bond in oxalate, but they also offer the prospect of providing new insights into the molecular processes that underpin the evolution of biological catalysts.

Microbial reduction of U(VI) at the solid-water interface; Jeon OH et al.; Microbial (Geobacter sulfurreducens) reduction of 0.1 mM U(VI) in the presence of synthetic Fe(III) oxides and natural Fe(III) oxide-containing solids was investigated in pH 6.8 artificial groundwater containing 10 mM NaHCO3. In most experiments, more than 95% of added U(VI) was sorbed to solids, so that U(VI) reduction was governed by reactions at the solid-water interface. The rate and extent of reduction of U(VI) associated with surfaces of synthetic Fe(III) oxides (hydrous ferric oxide, goethite, and hematite) was comparable to that observed during reduction of aqueous U(VI). In contrast, microbial reduction of U(VI) sorbed to several different natural Fe(III) oxide-containing solids was slower and less extensive compared to synthetic Fe(III) oxide systems. Addition of the electron shuttling agent anthraquinone-2,6-disulfonate (AQDS;

0.1 mM) enhanced the rate and extent of both Fe(III) and U(VI) reduction. These findings suggest that AQDS facilitated electron transfer from G. sulfurreducens to U(VI) associated with surface sites atwhich direct enzymatic reduction was kinetically limited. Our results demonstrate that association of U(VI) with diverse surface sites in natural soils and sediments has the potential to limit the rate and extent of microbial U(VI) reduction and thereby modulate the effectiveness of in situ U(VI) bioremediation.

A previously unexposed forest soil microbial community degrades high levels of the pollutant 2,4,6-trichlorophenol; Sanchez MA et al.; 2,4,6-trichlorophenol (2,4,6-TCP) is a hazardous pollutant that is efficiently degraded by some aerobic soil bacterial isolates under laboratory conditions. The degradation of this pollutant in soils and its effect on the soil microbial community are poorly understood. We report here the ability of a previously unexposed forest soil microbiota to degrade high levels of 2,4,6-TCP and describe the changes in the soil microbial community found by terminal restriction fragment length polymorphism (T-RFLP) analysis. After 30 days of incubation, about 50% degradation of this pollutant was observed in soils amended with 50 to 5,000 ppm of 2,4,6-TCP. The T-RFLP analysis showed that the soil bacterial community was essentially unchanged after exposure to up to 500 ppm of 2,4,6-TCP. However, a significant decrease in richness was found with 2,000 and 5,000 ppm of 2,4,6-TCP, even though the removal of this pollutant remained high. The introduction of Ralstonia eutropha JMP134 or R. eutropha MS1, two efficient 2,4,6-TCP degraders, to this soil did not improve degradation of this pollutant, supporting the significant bioremediation potential of this previously unexposed, endogenous forest soil microbial community.

Resistance of solid-phase U(VI) to microbial reduction during in situ bioremediation of uranium-contaminated groundwater; Ortiz-Bernad I et al.; Speciation of solid-phase uranium in uranium-contaminated subsurface sediments undergoing uranium bioremediation demonstrated that although microbial reduction of soluble U(VI) readily immobilized uranium as U(IV), a substantial portion of the U(VI) in the aquifer was strongly associated with the sediments and was not microbially reducible. These results have important implications for in situ uranium bioremediation strategies.

In situ expression of nifD in Geobacteraceae in subsurface sediments; Holmes DE et al.; In order to determine whether the metabolic state of Geobacteraceae involved in bioremediation of subsurface sediments might be inferred from levels of mRNA for key genes, in situ expression of nifD, a highly conserved gene involved in nitrogen fixation, was investigated. When Geobacter sulfurreducens was grown without a source of fixed nitrogen in chemostats with acetate provided as the limiting electron donor and Fe(III) as the electron acceptor, levels of nifD transcripts were 4 to 5 orders of magnitude higher than in chemostat cultures provided with ammonium. In contrast, the number of transcripts of recA and the 16S rRNA gene were slightly lower in the absence of ammonium. The addition of acetate to organic- and nitrogen-poor subsurface sediments stimulated the growth of Geobacteraceae and Fe(III) reduction, as well as the expression of nifD in Geobacteraceae. Levels of nifD transcripts in Geobacteraceae decreased more than 100-fold within 2 days after the addition of 100 microM ammonium, while levels of recA and total bacterial 16S rRNA in Geobacteraceae remained relatively constant. Ammonium amendments had no effect on rates of Fe(III) reduction in acetate-amended sediments or toluene degradation in petroleum-contaminated sediments, suggesting that other factors, such as the rate that Geobacteraceae could access Fe(III) oxides, limited Fe(III) reduction. These results demonstrate that it is possible to monitor one aspect of the in situ metabolic state of Geobacteraceae species in subsurface sediments via analysis of mRNA levels, which is the first step toward a more global analysis of in situ gene expression related to nutrient status and stress response during bioremediation by Geobacteraceae.

{Isolation and investigation of natural yeast strains resistant to heavy metal salts and radionuclides}; Enzymatic bioremediation: from enzyme discovery to applications; CSIRO Entomology, Canberra, Australian Capital Territory, Australia1. Enzymatic bioremediation is potentially a rapid method of removing environmental pesticide residues. Applications include the treatment of residues resulting from agricultural production and processing industries, such as the treatment of irrigation waters, surface-contaminated fruit and vegetables and spent dip liquors. 2. A specific application for some organophosphate-degrading enzymes involves detoxification of nerve agent stockpiles. Effective and affordable remediation requires highly specialized enzymes, so protein engineering techniques are being used to improve properties of various source enzymes to enhance catalytic rates, stability and substrate range. 3. Trials with an optimized organophosphate-degrading enzyme have shown the feasibility of such technology in various applications. 4. The enzymes developed for environmental remediation for specific pesticide classes also have applications as antidotes for high-dose pesticide poisonings and as prophylaxis for people at risk of high pesticide doses.

Modeling natural attenuation of chlorinated ethenes under spatially varying redox conditions; Widdowson MA; A three-dimensional model for the transport and reductive dechlorination of chlorinated ethenes in ground-water systems with variable redox conditions is demonstrated and applied to a pilot test for accelerated natural attenuation of trichloroethene (TCE). The rate and extent of biotransformation of TCE and chlorinated progeny is controlled by the dominant terminal electron accepting process (TEAP) that is simulated over space and time. The solute transport code, Sequential Electron Acceptor Model, 3D-transport, (SEAM3D) which simulates aerobic and sequential anaerobic biodegradation of organic carbon, is modified to implement the equations. Results of a generic model for TCE transport in ground-water systems with different redox conditions demonstrate that the degree of chlorinated ethene attenuation is influenced by background concentrations of aqueous- and solid-phase electron acceptors, but that model results are sensitive to other input parameters (inhibition coefficients, maximum rate of reductive dechlorination, biomass concentrations, and ground-water velocity). Simulation results of enhanced in situ bioremediation using dissolved organic carbon as a reducing agent show that spatial and temporal changes in the dominant TEAP and the subsequent rate of reductive dechlorination are adequately represented with the model. Initial concentrations of Fe(III) and the dechlorinating microbial population influence the simulated time lag observed during the pilot test.

Formulation of the CBC-model for modelling the contaminants and footprints in natural attenuation of BTEX; Maurer M et al.; This paper provides the details of the Coupled Biological and Chemical (CBC) model for representing in situ bioremediation of BTEX. The CBC model contains novel features that allow it to comprehensively track the footprints of BTEX bioremediation, even when the fate of those footprints is confounded by abiotic reactions and complex interactions among different kinds of microorganisms. To achieve this comprehensive tracking of all the footprints, the CBC model contains important new biological features and key abiotic reactions. The biological module of the CBC-model includes these important new aspects: (1) it separates BTEX fermentation from methanogenesis, (2) it explicitly includes biomass as a sink for electrons and carbon, (3) it has different growth rates for each biomass type, and (4) it includes inhibition of the different reactions by other electron acceptors and by sulfide toxicants. The chemical module of the CBC-model includes abiotic reactions that affect the footprints of the biological reactions. In particular, the chemical module describes the precipitation/dissolution of CaCO3, Fe2O3, FeS, FeS2, and S degrees. The kinetics for the precipitation/dissolution reactions follow the critical review in Maurer & Rittmann (2004).

Modeling intrinsic bioremediation for interpret observable biogeochemical footprints of BTEX biodegradation: the need for fermentation and abiotic chemical processes; Maurer M et al.; The intrinsic bioremediation of BTEX must be documented by the stoichiometric consumption and production of several other compounds, called 'footprints' of the biodegradation reaction. Although footprints of BTEX biodegradation are easy to identify from reaction stoichiometry, they can be confounded by the stepwise nature of the biodegradation reactions and by several abiotic chemical reactions that also produce or consume the footprints. In order to track the footprints for BTEX biodegradation, the following reactions need to be considered explicitly: (1) fermentation and methanogenesis as separate processes, (2) precipitation and dissolution of calcite, (3) precipitation and dissolution of amorphous iron monosulfide (FeS), (4) conversion of FeS into the thermodynamically stable pyrite (FeS2) with loss of sulfide and abiotic formation of H2, and (5) reductive dissolution of solid iron(III) by oxidation of sulfide. We critically review the research that underlies why these mechanisms must be included and how to describe them quantitatively. A companion manuscript develops and applies a mathematical model that includes these reactions.

Monitoring behaviour of catabolic genes and change of microbial community structures in seawater microcosms during aromatic compound degradation; Sei K et al.; The behaviour of microbial populations responsible for degradation of the aromatic compounds, phenol, benzoate, and salicylate, and changes of microbial community structures in seawater microcosms were analysed quantitatively and qualitatively using MPN-PCR and PCR-DGGE. The purpose of the study was to investigate the ecology of the entire microbial community during bioremediation. Bacterial populations possessing catechol 1,2-dioxygenase (C12O) DNA were evidently the primary degraders of phenol and benzoate, but others possessing catechol 2,3-dioxygenase (C23O) DNA increased to enhance substrate degradation under high-load conditions when the substrates were present for long periods. However, salicylate degradation was evidently facilitated by specific bacterial populations possessing C23O DNA. PCR-DGGE analyses suggested that bacterial populations already relatively dominant in the original microcosm contributed to phenol degradation. Bacteria composing a minor fraction of the original population apparently increased and contributed to benzoate degradation. Bacterial populations possessing C23O DNA were responsible for salicylate degradation, however, and different degrading bacteria were evidently selected for, depending on the initial salicylate concentration. Microbial community structure tended to be simplified by aromatic compound degradation. Thus, microbial monitoring can elucidate the behaviour of bacterial populations responsible for aromatic compound degradation and be used to assess the effects of bioremediation on intact microbial ecosystems.

Molecular diversity of tannic acid degrading bacteria isolated from tannery soil; Chowdhury SP et al.; AIMS: The aim of this study was to enrich and isolate bacteria from a tannery soil that were capable of utilizing tannic acid and gallic acid as sole source of carbon aerobically, and to characterize their diversity in order to identify efficient strains that can be used for tannin bioremediation. METHODS AND RESULTS: Bacterial strains were isolated after enrichment in minimal medium with tannic acid or gallic acid as sole carbon source. Polymerase chain reaction (PCR) restricted fragment length polymorphism of 16S rDNA {amplified ribosomal DNA restriction analysis (ARDRA)} and BOX-PCR was used to characterize their diversity. Two strains showing relatively high efficiency in degrading tannic acid and gallic acid were identified on the basis of carbon source utilization pattern (BIOLOG) and 16S rDNA sequence. CONCLUSIONS: Bacterial strains capable of degrading tannic acid and gallic acid could be grouped into six and seven clusters on the basis of ARDRA and BOX-PCR, respectively. i, b, i. On the basis of 16S rDNA sequence, the most efficient isolate degrading tannic acid belonged to Pseudomonas citronellolis, whereas the most efficient gallic acid degrader showed maximum phylogenetic relatedness to P. plecoglossicida. SIGNIFICANCE AND IMPACT OF THE STUDY: Aerobic tannic acid degraders such as the two strains isolated in this study can be used for tannin bioremediation, and in the study of genes involved in the production of tannase, an industrially important enzyme.

2,4,6-trinitrotoluene transformation by a tropical marine yeast, Yarrowia lipolytica NCIM 3589; Jain MR et al.; Yarrowia lipolytica NCIM 3589, a tropical marine degrader of hydrocarbons and triglycerides transformed 2,4,6-trinitrotoluene (TNT) very efficiently. Though this yeast could not utilize TNT as the sole carbon or nitrogen source, it was capable of reducing the nitro groups in TNT to aminodinitrotoluene (ADNT). In a complete medium containing glucose and ammonium sulphate as the available carbon and nitrogen sources respectively, the culture was able to completely transform 1 mM (227 ppm) of TNT under such conditions. A dual pathway was found to be functional, one of which resulted in the formation of the hydride-Meisenheimer complex (H(-)TNT) as a transiently accumulating metabolite that was subsequently denitrated to 2,4-dinitrotoluene (2,4-DNT), whereas the other pathway resulted in the formation of amino derivatives. The presence of increasing amounts of reducing equivalents in the form of glucose promoted better growth and the nitroreductases of this yeast to reduce the aromatic ring to 2,4-DNT although, the reduction of the nitro groups to amino groups was the major functional pathway. The ability of this tropical marine yeast to transform TNT into products such as 2,4-DNT which in turn could be metabolized by other microbes has implications in the use of this yeast for bioremediation of TNT polluted marine environments.

Response of the bacterial community to in situ bioremediation of organic-rich sediments; Vezzulli L et al.; A field trial experiment was carried out to assess the potential of bioremediation for mobilisation of carbon in organic-rich sediments. Both bioaugmentation (bio-fixed microorganisms) and biostimulation (oxygen release compounds—ORC) protocols have been tested and the response of the bacterial community has been described to assess the baseline for bioremediation potential. Multifactorial ANOVA revealed that bioaugmentation protocol had an effect in stimulate mobilisation processes and significantly enhanced extra-cellular enzymatic activity rates. In contrast biostimulation treatment did not have an effect on mobilisation rates but contributed to enhance bacterial efficiency through a maximization of the bacterial production:enzymatic activity ratio. Average calculation of net mobilised carbon showed that 23% increase of mobilised pool was accounted for bioaugmentation in summer. Although biostimulation accounted for a smaller increase in mobilised carbon (<10%), the use of ORC resulted in an increased mineralisation and net carbon loss via respiration. Based on our results, a conceptual model for application of bioremediation to face the problem of sediment eutrophication is discussed.

Bioremediation potential of a perchlorate-enriched sewage sludge consortium; Bardiya N et al.; The purpose of this work was to explore the reductive bioremediation potential of a perchlorate-enriched facultative anaerobic consortium. Rapid perchlorate reduction and bacterial growth were observed up to 1.84 g l(-1) of perchlorate, but not at 3.82 g l(-1) due to the toxicity. The specific growth rate of the mixed consortium was 0.1 h(-1). The consortium co-reduced perchlorate and nitrate with acetate as e- donor and carbon source. The presence of nitrate slowed down the perchlorate reduction rate. The other e- acceptors utilized include oxygen, chlorate, Cr(VI), and selenate. Over 95% of the 16 mg l(-1) of added Cr(VI) was reduced within 24 h of incubation with a high-density perchlorate-grown consortium. However, the consortium failed to couple growth with reduction of nitrite, sulfate, thiosulfate, and sulfite. During the search for autotrophic perchlorate reduction, many consortia from very diverse natural sources could not use sulfur compounds such as thiosulfate as e- donor.

Effect of root death and decay on dissipation of polycyclic aromatic hydrocarbons in the rhizosphere of yellow sweet clover and tall fescue; Parrish ZD et al.; A 12-mo greenhouse study was conducted to evaluate the contribution of root death and decay on the dissipation of polycyclic aromatic hydrocarbons (PAHs) in rhizosphere soil. The contaminated soil was previously treated by land-farming, but residual PAHs remained after treatment. Tall fescue (Festuca arundinacea Schreb.) and yellow sweet clover (Melilotus officinalis Lam.) were the target plants.

To specifically evaluate the effect of root decay on contaminant dissipation, plants were treated with glyphosate, a broad spectrum herbicide, to induce root decay. Although tall fescue treatments had the highest root and shoot biomass and root surface area, this plant did not result in the highest contaminant degradation rates. Significant differences were noted between treatments for seven PAHs, with the active yellow sweet clover resulting in 60 to 75% degradation of these compounds. Induced root death and decay did not produce a significant enhancement of PAH degradation. The PAH microbial degrader populations in the vegetated treatments were more than 100 times greater than those in the unvegetated control. The phospholipid fatty acid (PLFA) structural group profile shifted over the growing period, indicating a change in the community structure. In conclusion, phytoremediation was shown to be an effective polishing tool for PAH-affected soil previously subjected to biological treatment.

Ozonation of a landfill leachate: evaluation of toxicity removal and biodegradability improvement; Bila DM et al.; This work shows an evaluation of treatments for the leachate produced at the Gramacho Municipal Landfill in Rio de Janeiro state, Brazil. This leachate has very peculiar characteristics, with a high salinity level and very low biodegradability (BOD(5)/COD of 0.05). A sequence of processes was employed in the treatment of this leachate. Initially, a physicochemical treatment was used, while the second stage consisted of application of ozone to improve the biodegradability of the leachate. The final stage comprised a biological treatment. The physical-chemical treatment led to COD and DOC removal levels of 40 and 25%, respectively, with the use of Al(2)(SO(4))(3). The sequence of treatments proposed brought good results, with an increase in the BOD(5)/COD ratio from 0.05 to 0.3 after ozonation. The toxicity tests performed using Brachydanio rerio and Poecilia vivipara showed that the toxicity of the leachate had hardly been reduced by ozonation. These results are in agreement with the fact that, despite the higher BOD(5)/COD ratio, the biological process did not present a good performance. The total average removal levels of COD and DOC achieved using the combined treatment were 73 and 63%, respectively, for an ozone dose of 3.0gL(-1) by the leachate.

Effects of pH amendment on metal working fluid wastewater biological treatment using a defined bacterial consortium; van der Gast CJ et al.; The aim of this study was to determine whether pH amendment of a highly alkaline metal working fluid (MWF) wastewater would improve biological treatment in a bioreactor system following introduction of a bacterial inoculum (comprised of the following strains: Agrobacterium radiobacter, Comamonas testosteroni, Methylobacterium mesophilicum, Microbacterium esteraromaticum, and Microbacterium saperdae). The pH values tested were 6, 7, 8, and 9. Three replicate batch mode bioreactors inoculated with the bacterial inoculum (plus an abiotic control bioreactor) were operated for each of the four pH conditions. After 14 days, the final mean chemical oxygen demand (COD) reduction at pH 9 was 50 +/- 1.4%; at pH 8, 58 +/- 1.4%; pH 7, 65 +/- 1.0%; and pH 6, 75 +/- 2.7% of the initial COD (approximately 10,000 mg L(-1)), respectively. Interestingly, within 5 days, the pH in all inoculated bioreactors progressed toward pH 8. However, all abiotic control bioreactors remained at the pH at which they were amended. The fate of the inoculum, determined by denaturing gradient gel electrophoresis (DGGE) and by cluster analysis of the resulting DGGE profiles, revealed that the inocula survived throughout operation of all pH-amended bioreactors. Length-heterogeneity polymerase chain reaction (PCR) was used to track the population dynamics of individual strains. After 7 days of operation, M. esteraromaticum was the most abundant population in all bioreactors, regardless of pH. From our findings, it appears necessary to adjust the MWF wastewater from pH 9 to between 6 and 7, to achieve optimal biological treatment rates. (c) 2004 Wiley Periodicals, Inc.

From superior adaptation and function to brain dysfunction—the neglect of epigenetic factors; Saugstad LF; With optimal pregnancy conditions (natural, enriched diet which includes fish) African (Digo) infants are 3-4 weeks ahead of European/American infants in sensorimotor terms at birth, and during the first year. Infants of semi-aquatic sea-gypsies swim before they walk, and have superior visual acuity compared with us. With adverse pregnancy behaviour (fear of fat, a trend to dieting), neglecting the need for brain fat to secure normal brain development and function, we run a risk of dysfunction—death. Sudden Infant Death Syndrome victims have depressed birth weight, lower levels of marine fat in brainstem than controls, and >80 suffer multiple hypoxic episodes prior to death. Depressed birth weight (more than

10% below mean) is seen in learning and behaviour disorders, and a trend towards weights of less than 3kg is increasing, which supports a rise in antenatal sub optimality. Given marine fat deficiency in pregnancy and infancy, neurons starved for fuel could delay myelination and maturation in the latest developed Frontal Lobes.

The phylogenetic oldest Lateral Frontal Lobe System (feed-back mechanism etc.) derived from olfactory bulb-amygdala, which crosses in Anterior Commisure is probably spared, while the Medial Frontal Lobe System derived from Hippocampus-Cingulum and crosses in Corpus Callosum (delayed response task) is most likely affected.

The rise in infantile autism (intact vision and hearing) with deficit in delayed response task only, could suggest a deficit in the Medial Frontal Lobe System. The human species is unique; 70% of total energy to the foetus goes to development of the brain, which mainly consists of marine fat. It undergoes pervasive regressive events, before birth, in infancy and at puberty.

Minimal retraction of neuronal arborisation is advantageous. Attributable to adverse pregnancy childrearing practice, excessive retraction is likely prenatally and in infancy. Pubertal age affects the fundamental property of nervous tissue, excitability: excessive excitatory drive is seen in early, and a deficiency in late puberty.

It is postulated that with adequate marine fat, there is probably no risk of psychopathology at the extremes, whereas a deficiency could lead to paroxysmal (subcortical) dysfunction in early puberty, and breakdown of cortical circuitry and cognitive dysfunctions in late puberty. The post-pubertal psychoses, schizophrenia and manic-depressive psychosis at the extremes of the pubertal age continuum, with contrasting excitability and biological treatment, are probably the result of continuous dietary deficiency, which has inactivated the expression of genes for myelin development and oligodendrocyte-related genes in their production of myelin. The beneficial effect of marine fat in both disorders, in other CNS disorders as well as in developmental dyslexia (DD) and ADHD among others, supports our usual diet is persistently deficient. We have neglected the similarity of our great brain to other mammals, and our marine heritage. Given the amount of marine fat needed to secure normal brain development and function is not known, nor the present dietary level, it seems unduly conjectural to postulate that a dietary deficiency in marine fat is causing brain dysfunction and death. However, all observations point in the same direction: our diet focusing on protein mainly, is deficient, the deficiency is most pronounced in maternal nutrition and in infancy.

Potentials of biological oxidation processes for the treatment of spent sulfidic caustics containing thiols; Sipma J et al.; This research focused on the biological treatment of sulfidic spent caustics from refineries, which contain mainly hydrogen sulfide, methanethiol (MT) and ethanethiol (ET). Also various organic compounds can be present such as BTEX. Biological oxidation of 2.5 mM MT in batch experiments occurred after MT was first auto-oxidized into dimethyldisulfide (DMDS) whereafter oxidation into sulfate was completed in 350 h. DMDS as sole substrate was completely oxidized within 40 h. Therefore, DMDS formation seems to play an important role in detoxification of MT. Biological oxidation of ET and buthanethiol was not successful in batch experiments. Complete oxidation of MT and ET was observed in flow-through reactor experiments. Simultaneous oxidation of sulfide and MT was achieved when treating a synthetic spent caustic, containing 10 mM sulfide and 2.5 mM MT, in a bubble column reactor with carrier material at a hydraulic retention time of 6 h. Addition of 7.5 mM phenol, a common pollutant of spent caustics, did not adversely affect the biological oxidation process and phenol was completely removed from the effluent. Finally, three different spent caustics solutions from refineries were successfully treated.

Combined quantity management and biological treatment of sludge liquor at Hamburg's wastewater treatment plants—first experience in operation with the Store and Treat process; Laurich F; Store and Treat (SAT) is a new concept for the management of ammonium-rich process waste waters at wastewater treatment plants. It combines the advantages of quantity management and separate biological treatment, whereby both operations are carried out in the same tank. Now the first full-scale application of that method was realized in Hamburg. As first experience shows the process can help to increase nitrogen removal and to reduce energy consumption.

Electrochemical oxidation as a final treatment of synthetic tannery wastewater; Panizza M et al.; Vegetable tannery wastewaters contain high concentrations of organics and other chemicals that inhibit the activity of microorganisms during biological oxidations, so biorefractory organics that are not removed by biological treatment must be eliminated by a tertiary or advanced wastewater treatment. In this paper, the applicability of electrochemical oxidation as a tertiary treatment of a vegetable tannery wastewater was investigated by performing galvanostatic electrolysis using lead dioxide (Ti/PbO2) and mixed titanium and ruthenium oxide (Ti/TiRuO2) as anodes under different experimental conditions. The experimental results showed that both the electrodes performed complete mineralization of the wastewater. In particular, the oxidation took place on the PbO2 anode by direct electron transfer and indirect oxidation mediated by active chlorine, while it occurred on the Ti/TiRuO2 anode only by indirect oxidation. Furthermore, the Ti/PbO2 gave a somewhat higher oxidation rate than that observed for the Ti/TiRuO2 anode. Although the Ti/TiRuO2 required almost the same energy consumption for complete COD removal, it was more stable and did not release toxic ions, so it was the best candidate for industrial applications. With the Ti/TiRuO2 anode, the rate of tannery wastewater oxidation increased with the current density, pH, and temperature of the solution. These results strongly indicate that electrochemical methods can be applied effectively as a final treatment of vegetable tannery wastewater allowing the complete removal of COD, tannin, and ammonium and decolorization.

Biological nutrient removal model No.1 (BNRM1); Seco A et al.; This paper presents the results of the work carried out by the CALAGUA Group on Mathematical Modelling of Biological Treatment Processes: the Biological Nutrient Removal Model No.1. This model is based on a new concept for dynamic simulation of wastewater treatment plants: a unique model can be used to design, simulate and optimize the whole plant, as it includes most of the biological and physico-chemical processes taking place in all treatment operations. The physical processes included are: settling and clarification processes (flocculated settling, hindered settling and thickening), volatile fatty acids elutriation and gas-liquid transfer. The chemical interactions included comprise acid-base processes, where equilibrium conditions are assumed. The biological processes included are: organic matter, nitrogen and phosphorus removal; acidogenesis, acetogenesis and methanogenesis. Environmental conditions in each operation unit (aerobic, anoxic or anaerobic) will determine which bacterial groups can grow. Thus, only the model parameters related to bacterial groups able to grow in any of the operation units of a specific WWTP will require calibration. One of the most important advantages of this model is that no additional analysis with respect to ASM2d is required for wastewater characterization. Some applications of this model have also been briefly explained in this paper.

Comparison of different advanced oxidation process to reduce toxicity and mineralisation of tannery wastewater; Schrank SG et al.; Many organic compounds contained in wastewater are resistant to conventional chemical and/or biological treatment. Because of this reason different degradation techniques are studied as an alternative to biological and classical physico-chemical processes. Advanced Oxidation Processes (AOPs) probably have developed to become the best options in the near future. AOP while making use of different reaction systems, are all characterised by the same chemical feature: production of OH radicals (*OH). The versatility of AOPs is also enhanced by the fact that they offer different possibilities for OH radical production, thus allowing them to conform to specific treatment requirements. The main problem with AOPs is their high cost. The application of solar technologies to these processes could help to diminish that problem by reducing the energy consumption required for generating UV radiation. In this work, different AOPs (O3/ TiO2/UV, Fenton and H2O2/UV) were examined to treat tannery wastewater or as a pre-treatment step for improving the biodegradation of tannery wastewater, at different pH and dosage of the chemicals. Under certain circumstances retardation in biodegradation and/or an increase in toxicity may be observed within these treatment steps. Two different bioassays (Daphnia magna and Vibrio fischeri) have been used for testing the progress of toxicity during the treatment. In parallel other objectives were to analyse and identify organic compounds present in the untreated wastewater and arising degradation products in AOP treated wastewater samples. f, g, j, b. For this purpose substance specific techniques, e.g., gas chromatography-mass spectrometry (GC-MS) in positive electron impact (El(+)) mode and atmospheric pressure ionisation (API) in combination with flow injection analysis (FIA) or liquid chromatography-mass and tandem mass spectrometry (LC-MS or LC-MS-MS) were performed.

Short-term BOD (BODst) as a parameter for on-line monitoring of biological treatment process; Part II: instrumentation of integrated flow injection analysis (FIA) system for BODst estimation; Liu J et al.; An instrument with integrated flow injection analysis (FIA) system has been developed for on-line monitoring a process for conversion of biomass under field condition. The instrument consists of a newly designed biosensor for easy renewal of the bio-receptor without disassembling the sensor, a FIA controller for controlling the analysis operations, and a computer-based data acquisition system for data recording and processing. The instrument performed a sequence operations automatically including preparation of sample in the desired concentration, sample loading, sample injection, signal recording, data processing, and self-cleaning of the system. This makes the instrument being an interesting and promising device for on-line process monitoring.

Short-term BOD (BODst) as a parameter for on-line monitoring of biological treatment process. Part I. A novel design of BOD biosensor for easy renewal of bio-receptor; Liu J et al.; A novel design of a biochemical oxygen demand (BOD) biosensor has been developed for on-line monitoring of easily biodegradable organic compounds in aqueous samples. The biological recognition element of the sensor could be easily renewed by injecting new bacterial paste without disassembling the sensor system. The sensor measurements were carried out in the initial-rate mode using a flow injection (FI) system, resulting in 60 s for one sample analysis followed by a recovery time less than 10 min. The sensor performance achieved showed a wide detection linearity over the range of 5-700 mg BOD5.l(-1) and a generally good agreement between the BOD values estimated by the biosensor and the conventional 5-day test. Furthermore, the precision test was in the control range (i.e. repeatability < or = /+/-7.5%/, reproducibility < or = /+/-7.3%/). The sensor could be used over 1 week in continuous test, however, the best performance was found within the first 24 h where standard deviation of the sensor response was +/ -2.4%. The design of the sensor allows easy and fast renewal of the cells used as sensing elements. Replacement of biological recognition element and calibration of the sensor responses can be performed in a rather simple procedure on a daily regular basis. By using a mixed culture as the bio-receptor, one gets a sensor that reacts to a wide range of substrates. The new sensor construction will thus allow fast and convenient replacement of the bio-receptor and on-line assay of a broad range of substrates. This makes the sensor being an interesting and promising candidate for on-line monitoring of biological treatment process.

Chemical characterisation of natural organic substrates for biological mitigation of acid mine drainage; Gibert O et al.; The current approach of the biological treatment of acid mine drainage by means of a passive remediation system involves the choice of an appropriate organic substrate as electron donor for sulphate reducers. Nowadays this selection is one of the critical steps in the performance of such treatment, as this depends to a great extent on the degradability of the organic substrate. Thus, a prior characterisation of the organic substrate predicting its biodegradability would be desirable before embarking on an extensive large-scale application. The aim of this study was to correlate the chemical composition (lignin content) of four different natural organic substrates (compost, sheep and poultry manures, oak leaf) and their capacity to sustain bacterial activity in an attempt to predict biodegradation from chemical characterisation. The results showed that the lower the content of lignin in the organic substrate, the higher its biodegradability and capacity for developing bacterial activity.

Of the four organic materials, sheep and poultry manures and oak leaf evolved reducing conditions and sustained active sulphidogenesis, which coupled with the decrease in sulphate concentration indicated bacterial activity. Sheep manure was clearly the most successful organic material as electron donor (sulphate removal >99%), followed by poultry manure and oak leaf (sulphate removal of 80%). Compost appeared to be too poor in carbon to promote sulphate-reducing bacteria activity by itself. Column experiments emphasised the importance of considering the residence time as a key factor in the performance of continuous systems. With a residence time of 0.73 days, sheep manure did not promote sulphidogenesis. However, extending residence time to 2.4 and 9.0 days resulted in an increase in the sulphate removal to 18% and 27%, respectively.

Landfill leachate treatment by yeast and bacteria based membrane bioreactors; Wichitsathian B et al.; Biological treatment of medium-age landfill leachate was investigated on a membrane bioreactor. The experiments were conducted in two 5-L reactors with immersed hollow fiber microfiltration membranes. One reactor was operated with a mixed bacterial culture termed as bacteria based membrane bioreactor (BMBR) while the other with mixed yeast culture termed as yeast based membrane bioreactor (YMBR). The leachate was characterized with a chemical oxygen demand (COD) concentration of 7000-9000mg/L, biochemical oxygen demand (5 days) to chemical oxygen demand ratio (BOD5/COD) of 0.35-0.45 and total Kjeldahl nitrogen (TKN) of 1800-2000mg/L. The performance was assessed with and without ammonia stripping. In both the reactors, the average COD and TKN removal efficiency without ammonia stripping ranged between 52-66 and 14-28%, respectively. The performance of both the membrane bioreactors improved with ammonia stripping in terms of both COD (72-76%) and TKN (82-89%) removal efficiency. Though, the difference in the performance of the BMBR and YMBR was not significant in terms of COD removal, the YMBR showed better removal efficiency in terms of BOD5. The molecular weight cut-off showed that the degradation pathway of the leachate by bacterial and yeast are different. In regard to membrane fouling, the YMBR showed better performance with lower trans-membrane pressure as well as longer operating time. This superior performance of the YMBR could be due to the structure of yeast cells which are larger in size as well as reduced soluble extracellular polymeric substances (EPS) production, which are the main cause of membrane biofouling.

Membrane bioreactors for the removal of anionic micropollutants from drinking water; Crespo JG et al.; Biological treatment processes allow for the effective elimination of anionic micropollutants from drinking water. However, special technologies have to be implemented to eliminate the target pollutants without changing water quality, either by adding new pollutants or removing essential water components. Some innovative technologies that combine the use of membranes with the biological degradation of ionic micropollutants in order to minimize the secondary contamination of treated water include pressure-driven membrane bioreactors, gas-transfer membrane bioreactors and ion exchange membrane bioreactors.

Water Sci Technol, 2004, 50(3), 195 - 206

Biological treatment of whitewater in a laboratory process in order to determine kinetic parameters for model development; Alexandersson T et al.; Implementation of an in-mill biological treatment plant is one solution to the problems associated with closure of whitewater systems. It is, however, important to operate the treatment with low concentration of nutrients in the effluent.

The effect on the COD reduction from decreased additions of NH4-N and PO4-P were investigated in three parallel aerobic suspended carrier reactors during treatment at 46 to 48 degrees C of whitewater from a recycled paper mill producing liner and fluting. In the reference reactor, a COD reduction of 89% was achieved and 45.6 mg NH4-N/(g COD reduced) and 11.6 mg PO4-P/(g COD reduced) was consumed at an organic load around 20 kg COD/(m3 x d). Reduced additions of NH4-N decreased the COD reduction. Addition of 56% of the consumption of NH4-N in the reference reactor resulted in a COD reduction of 80%. The response from decreased addition of PO4-P was different compared to NH4-N but it could not be determined if this is due to unsuitable experimental design or a different reaction mechanism. Reducing the addition of PO4-P to 26% of the consumption of PO4-P in the reference reactor, decreased the COD reduction to 83%. The main conclusion from the experiment is: biological treatment has the potential of treating whitewater from recycled paper mills with low effluent nutrient concentrations.

Pulp mill process closure: a review of global technology developments and mill experiences in the 1990s; Stratton SC et al.; The impact of effluent discharges continues to be an important issue for the pulp manufacturing industry. Considerable progress has been made in pollution prevention to minimize waste generation, so-called manufacturing "process closure." Since the mid-1980s many important technologies have been developed and implemented, many of these in response to organochlorine concerns. Zero effluent operation is now a reality for a few bleached chemi-thermomechanical pulp (BCTMP) pulp mills.

In kraft pulp manufacturing, important developments include widespread adoption of new cooking techniques, oxygen delignification, closed screening, improved process control, new bleaching methods, and systems that minimize pulping liquor losses. Coupled to this is a commitment to reduce water use and maximize reuse of in-mill process streams. Some companies pursued bleach plant closure, and many have been successful in eliminating a portion of their bleaching wastewaters. However, the difficulties inherent in closing bleach plants are considerable. For many mills the optimal solution has been found to be a high degree of closure coupled with external biological treatment of the remaining process effluent. No bleach plants at papergrade bleached kraft mills are known to be operating effluent-free on a continuous basis. This paper reviews the important worldwide technological developments and mill experiences in the 1990s that were focused on minimizing environmental impacts of pulp manufacturing operations.

Integrated biological treatment of recalcitrant effluents from pulp mills; Ortega-Clemente A et al.; This work aimed at determining the degree of depuration of a recalcitrant effluent (weak black liquor, WBL) achieved in a series treatment consisting of a first stage methanogenic fluidised bed reactor followed by a second stage aerobic, upflow reactor packed with "biocubes" of Trametes versicolor immobilised onto small cubes of holm oak wood. The mesophilic, lab scale methanogenic fluidised bed reactor contained a microbial consortium immobilised onto granular activated carbon 500 microm average size. The process removed decreasing amounts of organic matter at decreasing hydraulic retention times (HRT), eventually reaching an average of 50% at 0.5 day HRT. Colour and ligninoid removals also decreased with decreasing HRT. Although the methanogenic fluidised bed reactor provided an effective treatment for the degradable organic matter, important concentrations of recalcitrant organic matter and colour still remained in the anaerobic effluent.

This anaerobic effluent was fed to the aerobic packed bed reactor. Two HRT were tested in this unit, namely 5 and 2.5 days. The reactor averaged an organic matter removal in the range of 32% COD basis, during an experimental run of 95 days. Colour and ligninoid contents were removed in high percentages (69% and 54%, respectively). There was no significant difference in reactor performance at 5- and 2.5-day HRT. There was a positive correlation between pollutant removal efficiencies and Laccase activity in crude extracts of the reactor liquor. No supplemental soluble carbohydrate was required to sustain the fungus activity and the consistent reactor performance. Overall, the two-stage treatment achieved approximately a 78% removal of the original organic matter of the WBL (COD basis) and ca. 75% of colour and ligninoid contents.

{Treatment of coloured industrial effluents with Pleurotus spp}; Rodriguez S et al.; The decolouration of fermentation residues (vinasse) and liquid extract of coffee pulp by the mushroom Pleurotus ostreatus was studied in addition to laccase activity. The fungus was inhibited in both residues when they remained undiluted. In submerged cultivation on wastewaters a good production of biomass (14.8 g/l for vinasse and 5.4 g/l for extract of coffee pulp) and also laccase activity (14.1 U/ml for vinasse and 3.0 U/ml for extract of coffee pulp) up to the 10 days of fermentation was observed, being significantly greater in the culture with vinasse. It was shown that treatment with this mushroom reduces both the chemical oxygen demand and the colour, contributing to their biological treatment.

Biostabilization of municipal solid waste; Adani F et al.; A mechanical-biological process for municipal solid waste (MSW) treatment was monitored for one year. Mechanical pre-treatment provided two fractions. The oversize fraction (diameter > 50 mm) (yield of 600 g kg(-1) ww) (46 Mg day(-1)) was used for refuse derived fuel production, after undergoing a mechanical refining processes, because of low moisture content (200-250 g kg(-1)) and high calorific value (2500-2800 kcal kg ww(-1)). The undersize fraction (diameter < 50 mm) (yield 400 g kg(-1) ww) (30 Mg day(-1)) contained about 800 g kg(-1) of the MSW organic matter. This fraction was biologically treated using an aerobic process with an organic waste fraction from separate collection (77 Mg day(-1)) and recycled stabilized material (62 Mg day(-1)) obtained from end-product sieve (diameter < 20 mm) used as bulking agent. A retention time of three weeks was sufficient to obtain stabilized products in agreement with up-dated rules of the Lombardy Region (North Italy) regarding biostabilization and composting processes. Dynamic Respiration Index (DRI), such as required by both Lombardy Region rules and suggested by the European Community, was chosen in preference to other indices in order to assess the degree of biological stability of the end products. A mean DRI value of 1164 mg O2 kg SV(-1) h(-1) was obtained and is in agreement with the proposed limit of 1000+/-200 mg O2 kg SV(-1) h(-1). Self-heating test, potential biogas production and fermentable volatile solids were also used as parameters to describe the potential impact of treated waste, providing further useful information. Nevertheless, all of these methods revealed analytical or interpretative limits. A complete mass balance of the biological treatment section showed that, from a net input of 107 Mg day(-1), only 250 g kg(-1) (27 Mg day(-1)) of the waste needed to be landfilled, with 750 g kg(-1) (80 Mg day(-1)) being lost as CO2 and H2O.

Metal solubilization from metal-containing solid materials by cyanogenic Chromobacterium violaceum; Faramarzi MA et al.; Different cyanogenic bacterial strains (Chromobacterium violaceum, Pseudomonas fluorescens, Bacillus megaterium) were cultivated under cyanide-forming conditions in the presence of metal-containing solids such as nickel powder or electronic scrap. All microorganisms were able to form water-soluble metal cyanides, however, with different efficiencies. C. violaceum was able to mobilize nickel as tetracyanonickelate {Ni(CN)4(2-)} from fine-grained nickel powder. Gold was microbially solubilized as dicyanaoaurate {Au(CN)2-} from electronic waste. Additionally, cyanide-complexed copper was detected during biological treatment of shredded printed circuit boards scrap. Regarding the formation of tetracyanonickelate, C. violaceum was more effective than P. fluorescens or B. megaterium. Besides a few previous reports on gold solubilization from gold-containing ores or native gold by C. violaceum, the findings demonstrate for the first time the microbial mobilization of metals other than gold from solid materials and represent a novel type of microbial metal mobilization based on the ability of certain microbes to form HCN. The results might have the potential for industrial applications (biorecovery, bioremediation) regarding the treatment of metal-containing solids since metal cyanides can easily be separated by chromatographic means and be recovered by sorption onto activated carbon.

How to avoid pharmaceuticals in the aquatic environment; Larsen TA et al.; Pharmaceuticals and other micropollutants in wastewater pose a new challenge to wastewater professionals as well as to the pharmaceutical industry. Although there is a great deal of uncertainty concerning the possible detrimental effects on the aquatic ecosystems, the precautionary principle—or possibly new scientific evidence—may give rise to more stringent demands on wastewater treatment in the future. In conventional wastewater treatment plants, a combination of biological treatment with high sludge residence times and ozonation of the effluent seems to be the most promising technology. Ozonation, however, is an energy-intensive technology. Moreover, in conventional end-of-pipe systems a large part of the pollutants will always be lost to the environment due to leaking, primarily during rain. In the long term, source separation offers the more sustainable solution to the entire wastewater problem, including organic micropollutants. Urine source separation is an elegant solution to the problems of nutrients and pharmaceuticals alike and losses of untreated pollutants to the environment can be minimized. Although few technologies for the separate treatment of urine have been developed to date, the 100-500 times higher concentrations of micropollutants promise more efficient conditions for all removal technologies known from conventional wastewater treatment.

Fate and stability of 14C-labeled 2,4,6-trinitrotoluene in contaminated soil following microbial bioremediation processes; Weiss M et al.; Biological treatment of 2,4,6-trinitrotoluene (TNT) in soil rarely results in complete mineralization of the parent compound. More often, the largest proportion of the TNT carbon is incorporated into the soil organic matrix. Therefore, we evaluated the stability of nonextractable residues from various bioremediation processes of 14C-TNT in soils. The extractable amounts of the residual radioactivity varied between 7 and 33% and thus the nonextractable amount between 93 and 67% (3-15% in fulvic acids, 26-46% in humic acids, and 27-44% in the humin fraction). The residue-containing soils were analyzed for the release of radioactivity after treatment by physical (freeze and thaw, grinding of soil, and steam extraction), chemical (acid rain and addition of metal complexing agent), and biological methods (addition of compost, white rot fungi, radical-generating enzymes, and germination of plants). Freeze and thaw treatment and grinding of the soil did not alter the partitioning of the label significantly. Steam extraction and acid rain extraction increased the water extractability to 11 to 29% and to 51.6% in the native TNT-contaminated soil. The addition of ethylenediamine-tetraacetate (EDTA) increased the extractability from 7 to 12%. d, h, a, c, j, b. After biological treatment, only slightly increased extractability (<10%) was observed. No increase of extractable TNT or known metabolites was observed with any of the treatments. Thus, under the treatment conditions applied in this study, the residues formed during microbial transformation of TNT may be biogenic residues with low mobilization potential and low hazardous impact.

Biotreatment of wastewater using aquatic invertebrates, Daphnia magna and Paramecium caudatum; Shiny KJ et al.; A number of major changes have occurred over the past few years, which give cause for a re-examination of conventional wastewater treatment methods. Among these are growing problems of worldwide energy and food shortages and nutrients not removed by conventional secondary processes causing algal blooms and other problems in the receiving waters. The global increase in wastewater calls for innovative low cost technology approaches to its recycling. Biotreatment systems, utilizing living organisms are receiving growing attention since they are ecologically sound, cheap and applicable in areas without land constraints. Filter feeders (both invertebrates and vertebrates) are promising in this area since they can remove suspended organic matter and bacteria, even in the size range of microns. In the present study biological treatment of municipal wastewater using two invertebrates—Paramecium caudatum, a protozoan and Daphnia magna, a cladoceran was investigated. Analysis at pre-experimental and post-experimental stages revealed the potential of these species in abatement of water pollution. D. magna was more efficient than P. caudatum in laboratory-scale studies.

Fe(III)-EDTA complex abatement using a catechol driven Fenton reaction combined with a biological treatment; Oviedo C et al.; A combined chemical oxidation (catechol-driven Fenton reaction) followed by a biological treatment was used to degrade Fe(III)-EDTA (1.34 mM). The chemical treatment was inspired in fungal non-enzymatic wood rot mechanisms that use dihydroxybenzens in order to promote the Fenton reaction to breakdown wood structures. This chemical pre-oxidation used catechol (50 microM) and H202 (20 mM) and the reaction products were identified by GC-MS. In addition, a biological treatment was coupled using the baker's yeast Saccharomyces cerevisiae. The combined chemical biological treatment achieved 100% EDTA degradation, 68 % total organic carbon removal and 90% iron removal.

The World Federation of Societies of Biological Psychiatry (WFSBP) guidelines for the biological treatment of bipolar disorders, part III: maintenance treatment; Grunze H et al.; As with the two preceding guidelines of this series, these practice guidelines for the pharmacological maintenance treatment of bipolar disorder were developed by an international task force of the World Federation of Societies of Biological Psychiatry (WFSBP). Their purpose is to supply a systematic overview of all scientific evidence relating to maintenance treatment. The data used for these guidelines were extracted from a MEDLINE and EMBASE search, from recent proceedings from key conferences and various national and international treatment guidelines. The scientific justification of support for particular treatments was categorised into four levels of evidence (A-D). As these guidelines are intended for clinical use, the scientific evidence was not only graded, but also reviewed by the experts of the task force to ensure practicality.

Evaluation of virological quality of sewage from four biological treatment plants by a nested-PCR technique; Komninou G et al.; In order to determine the virological quality of sewage from four biological treatment plants in Greece (two in the city of Athens and two in the city of Patras), 92 raw sewage samples were analysed for the presence of enteroviruses and adenoviruses during the period from October 2000 to February 2003. A nested-PCR method was used in order to increase the sensitivity of virus detection. Enteroviruses were detected in 43 samples (47%) and adenoviruses in 75 samples (81.5%) of raw sewage by nested PCR. The more frequent isolation of adenoviruses in raw sewage indicated their stability as virological indicators of the pollution of the environment and their increased persistence in sewage.

Hydration properties of eco-cement pastes from waste sludge ash clinkers; Lin KL et al.; Three types of hydraulic cements have been developed by incorporating sludge ash from a primary sewage treatment plant and a water purification plant, as well as slag from steelworks (ferrate), as a partial replacement for clay, silica, alumina, and iron oxide in raw cement meal. The raw meal for the pre-determined recipes was prepared by heating it to 1400 degrees C for 6 hr in a clinkerization process, using a simulated incinerator and smelter. The major components of ordinary Portland cement, C3S, C2S, C3A, and C4AF, were all found in the clinkers. Of the three types of eco-cements, the eco-cement A paste was most similar to ordinary Portland cement in terms of composition and compressive strength development, while the eco-cement B paste showed early strength development. The differential thermal analysis species analyses indicated that the hydrates in the eco-cement pastes were mainly calcium hydroxide and CSH gels, like those found in ordinary Portland cement paste. Moreover, the degree of hydration, as determined by nuclear magnetic resonance, increased in all eco-cement pastes with an increasing curing age. The results indicate that it indeed is feasible to use sludge ash and ferrate to replace up to 20% of the mineral components of raw materials for cement.

Aquatic ecosystem as a bioreactor: water purification and some other functions; Ostroumov SA; A fundamental concept is proposed of aquatic ecosystem as a bioreactor that carries out the function of water purification in natural water bodies and streams. The ecosystem as a bioreactor has the following characteristic attributes: (1) it is a large-scale (large-volume) bioreactor; (2) it is a diversified (in terms of the number of taxa and the scope of functional activities) bioreactor; (3) it possesses a broad range of biocatalytic (chemical-transforming and degrading) capabilities. New experimental data on xenobiotics-induced inhibition of the functioning of the molluscs Unio tumidus, U. pictorum, M. galloprovincialis and Limnaea stagnalis emphasize the potential ecological hazard from sublethal concentrations of pollutants (including those exemplified by synthetic surfactants and detergents).

Detection of enteric viruses and bacterial indicators in German environmental waters; Pusch D et al.; A German mining lake and the supplying surface waters, which are located downstream of a sewage plant, were examined regarding their microbiological and virological quality. Between October 2002 and September 2003, specific PCR methods were used to determine the occurrence of enteric viruses in 123 water specimens drawn at different sites downstream of the waste water treatment plant and in 9 samples from the sewage plant influent. Detection rates in sewage plant effluents and surface water samples depended on sampling sites and were: 29-76% for enterovirus (EntV), 24-42% (astrovirus, AstV), 15-53% (norovirus, NV), 3-24% (rotavirus, RoV), 5-20% (hepatitis A virus, HAV) and 20% (adenovirus, AdV). AstV genome load of selected samples was between 3.7 x 10(3) to 1.2 x 10(8) genome equivalents per liter (gen.equ./l), depending on sampling location; NV average genome load ranged from 1.8 x 10(4) to 9.7 x 10(5) gen.equ./l. Cell culture methods showed that three out of 18 PCR positive samples contained infectious EntV. Even though microbiolical parameters such as Escherichia coli, enterococci and coliphages indicated acceptable microbiological water quality, the virological data of this study suggest the possibility that surface waters may be a source for enteric viral infections.

Studies on the effect of humic acids and phenol on adsorption-ultrafiltration process performance; Mozia S et al.; Application of pressure-driven membrane processes, such as ultrafiltration (UF) and microfiltration (MF) for surface water treatment have become very popular during last decades. Membrane fouling by humic substances (HS) is one of the major limiting factors in these processes. In order to alleviate the unfavorable effects of the presence of HS in the feed on the process performance UF and MF are often combined with adsorption on powdered activated carbon (PAC). The main goal of the present study was to evaluate the effect of humic acid (HA) on membrane fouling during UF. Moreover, the effect of PAC addition to the feed on UF process, especially on flux decline was determined. The applicability of the adsorption-ultrafiltration (PAC/UF) system to purification of water containing low (phenol) and high molecular (HA) was also investigated. Three different polymer UF membranes, prepared from polysulfone (PSF), cellulose acetate (CA) or polyacrylonitrile (PAN) were applied. It was found that the membranes prepared from PSF and CA are very susceptible to fouling caused by HA. The permeate flux decreased for ca. 50% during UF of HA solution through the PSF membrane and for ca. 45%-through the CA membrane. In the case of the PAN membrane, a negligible effect of HA on the flux was observed. On the basis of the FTIR spectra it was found that the drop in the permeate flux through these membranes may result from interactions between the negatively charged functional groups present on the membrane surface, such as carboxyl groups (CA) and sulfone groups (PSF) with HA, which results in coating of the membrane surface with HA. When PAC was added to the feed containing HA, the permeate flux through the CA and PAN membranes was maintained on a practically unchanged level. However, in case of the PSF membrane, a 50% drop in the permeate flux in comparison with the flux value, when process was conducted without PAC addition was observed. That was supposed to be due to attractive forces among hydrophobic PAC particles, HA molecules and PSF membrane surface. The performed studies showed that the application of PAC/UF system was very effective in the removal of organic substances having both, low and high molecular weights. The role of PAC suspended in a feed in the PAC/UF system is the adsorption of low molecular organic compounds, which cannot be removed by UF alone.

Optimization of the extended terminal subfluidization wash (ETSW) filter backwashing procedure; Amburgey JE; The increased passage of particles and microorganisms through granular media filters immediately following backwashing is a common problem known to the water treatment community as filter "ripening" or maturation. While several strategies have been developed over the years to reduce the impact of this vulnerable period of the filtration cycle on finished water quality, this research involves a recently developed filter backwashing strategy called the extended terminal subfluidization wash (ETSW). ETSW is a method of terminating the backwash cycle with a subfluidization wash for a period of time sufficient to pass one theoretical filter-volume of water upward through the filter. ETSW was shown to remove significantly greater quantities of backwash remnant particles thereby reducing the magnitude of filter ripening turbidity and particle count spikes. Optimum ETSW flow rates were determined for deep-bed anthracite and granular activated carbon filters herein by monitoring filter effluent turbidities and particle counts during the filter ripening period. Optimality of the coagulation process was also shown to influence the magnitude of filter ripening particle passage. ETSW was found to be equally effective for biological and conventional deep-bed anthracite filters.

An improvement on the power law for the description of particle size distributions in potable water treatment; Ceronio AD et al.; In potable water treatment, the use of the power law to describe particle size distributions (PSDs) in particle counting practice is common. The power law is popular because it allows the reduction of numerous data bits to two meaningful parameters that completely describe the size distribution characteristics of a particle suspension. The model is however flawed. This paper presents the further development of an improved model (the variable-beta model) first proposed by Lawler (1997). Both the power law model and variable-beta model are used to describe the PSDs of a large number of potable water treatment samples taken from full-scale plants and the resulting correlations are compared. The findings from the comparison of data reduction methodologies support the argument that the variable-beta model is fundamentally more correct than the power law model and consequently describes the PSDs better.

Epidemiology of Cryptosporidium molnari in Spanish Gilthead Sea Bream (Sparus aurata L.) and European Sea Bass (Dicentrarchus labrax L.) Cultures: from Hatchery to Market Size; Sitja-Bobadilla A et al.; A long-term epidemiological study of Cryptosporidium molnari in aquacultured European sea bass (ESB) and gilthead sea bream (GSB) was performed in different types of facilities on the Atlantic, Cantabric, and Mediterranean coasts. Four types of studies were carried out. In study A, fish raised from juveniles to marketable size (ongrowing stage) were periodically sampled in three different types of cultures. Studies B and C focused on hatchery and nursery facilities. In study D, occasional samplings were performed during mortality or morbidity outbreaks. As a general trend, C. molnari was more prevalent in GSB than in ESB. Data on the distribution pattern of C. molnari in total sampled GSB (studies A, B, and D) had a variance higher than the mean (overdispersion). In GSB (study A), the type of ongrowing system (sea cages, earth ponds, or indoor tanks) was found to have no significant effect. There was a significant relationship between the presence of the parasite and both fish weight and season. The highest infection values were recorded in spring. Prevalence and intensity had convex weight profiles, with a peak in 30- to 100-g fish. In study D, the prevalence of infection was higher in fish recently introduced in sea cages and in preongrowing systems. In studies B and C, fish were almost never infected before entering the postlarval and nursery facilities. The parasite seems to enter the host mainly through the water in production steps with less stringent water treatment. Recirculation systems and fish cannibalism could contribute to oocyst concentration and dispersion in aquaculture facilities.

Melatonin improved the disturbances in hepatic prooxidant and antioxidant balance and hepatotoxicity induced by a high cholesterol diet in C57BL/6J mice; Balkan J et al.; We examined the effect of melatonin in prooxidant and antioxidant state in the liver of C57BL/6J mice fed on a high cholesterol (HC) diet. Mice were fed with normal mice chow containing 1.5% cholesterol and 0.5% cholic acid for 4 months without and with melatonin (10 mg/L in drinking water) treatment. HC diet was observed to increase malondialdehyde (MDA) and diene conjugate (DC) levels in the liver. This diet lowered glutathione (GSH), alpha-tocopherol, and total ascorbic acid levels as well as glutathione peroxidase (GSH-Px) and glutathione transferase (GST) activities in the liver, but hepatic superoxide dismutase (SOD) activity remained unchanged. Although melatonin treatment did not affect these parameters in mice fed a normal diet, it reduced hepatic MDA and DC levels in mice fed an HC diet. Hepatic alpha-tocopherol and ascorbic acid levels increased, but hepatic GSH levels remained unchanged in the melatonin-treated HC group as compared to the HC group. Melatonin treatment was found to increase liver GSH-Px and GST activities in mice fed an HC diet. However, SOD activity did not alter in the liver of hypercholesterolemic mice following melatonin treatment. In addition, the histopathological lesions observed in the cholesterol-plus-melatonin group were less severe than those seen in the cholesterol group. According to these observations, we can say that melatonin treatment has an ameliorating effect on the disturbances in prooxidant and antioxidant balance and histopathological lesions in the liver of mice following cholesterol feeding.

Comparative study of different surface decontaminants on chicken quality; Sinhamahapatra M et al.; (1) A comparative study on the effect of different surface decontaminants: hot water at 70 degrees C for one minute; 2% lactic acid for 30 s; 1200 p.p.m. acidified sodium chlorite (ASC) solution for 5 s and 50 p.p.m. chlorine solution for 5 min in the form of dips and sprays on the surface of dressed broilers for 0, 24 and 48 h of storage was conducted. (2) The variables studied were, total plate count (TPC), presumptive coliform count (PCC), pH and extract release volume (ERV). All treatments reduced TPC and PCC. (3) Lactic acid dip and hot water dip were the most effective for reducing TPC (1.36 and 1.28 log/cm2, respectively) with no significant difference between them. (4) ASC and hot water in dip could diminish PCC (1.37 and 1.34 log/cm2, respectively) and did not vary significantly. (5) No treatment affected muscle pH, water holding capacity (WHC), ERV, appearance, smell, tenderness and overall acceptability of treated broilers significantly. (6) Hot water treatment is the cheapest, most convenient and simplest decontamination technique for hygienic and wholesome poultry production.

{Relationship between phosphorus and bacterial regrowth in drinking water}; Jiang DL et al.; Limitation of phosphorus on bacterial regrowth was studied in source water, water treatment process and a main distribution pipe of J waterwork in T city, by modified assimilable organic carbon (AOC) method and microbially available phosphorus (MAP) analysis. Based on the study, the result showed that: (1) Concentration of MAP was higher in source water and water treatment process, which was 5-38 microg/L (PO4(3-)-P). However in water distribution system, concentration of MAP was lower, which was less than 5 microg/L (PO4(3-)-P). It changed very little in distribution system. (2) Traditional treatment process can remove MAP efficiently 34.0%-83.7% of MAP can be removed in the process. (3) Generally in source water and water treatment process, there was no evident difference between AOC(potential), AOC(P) and AOC(native). So AOC was the most important factor of bacterial growth. In the main distribution pipe, the concentration of AOC(potential) and AOC(P) were 2-8.7 times of the concentration of AOC(native). So phosphorus limited bacterial regrowth.

Performance of a constructed wetland treating intensive shrimp aquaculture wastewater under high hydraulic loading rate; Lin YF et al.; A water treatment unit, mainly consisting of free water surface (FWS) and subsurface flow (SF) constructed wetland cells, was integrated into a commercial-scale recirculating aquaculture system for intensive shrimp culture. This study investigated performance of the treatment wetlands for controlling water quality. The results showed that the FWS-SF cells effectively removed total suspended solids (55-66%), 5-day biochemical oxygen demand (37-54%), total ammonia

(64-66%) and nitrite (83-94%) from the recirculating water under high hydraulic loading rates (1.57-1.95m/day). This led to a water quality that was suitable for shrimp culture and effluent that always satisfied the discharge standards. The area ratios of wetlands to culture tank being demonstrated (0.43) and calculated (0.096) in this study were both significantly lower than the reported values. Accordingly, a constructed wetland was technically and economically feasible for managing water quality of an intensive aquaculture system.

Water, sanitation, and hygiene interventions to reduce diarrhoea in less developed countries: a systematic review and meta-analysis; Fewtrell L et al.; Many studies have reported the results of interventions to reduce illness through improvements in drinking water, sanitation facilities, and hygiene practices in less developed countries. There has, however, been no formal systematic review and meta-analysis comparing the evidence of the relative effectiveness of these interventions. We developed a comprehensive search strategy designed to identify all peer-reviewed articles, in any language, that presented water, sanitation, or hygiene interventions. We examined only those articles with specific measurement of diarrhoea morbidity as a health outcome in non-outbreak conditions. We screened the titles and, where necessary, the abstracts of 2120 publications. 46 studies were judged to contain relevant evidence and were reviewed in detail. Data were extracted from these studies and pooled by meta-analysis to provide summary estimates of the effectiveness of each type of intervention. All of the interventions studied were found to reduce significantly the risks of diarrhoeal illness. Most of the interventions had a similar degree of impact on diarrhoeal illness, with the relative risk estimates from the overall meta-analyses ranging between 0.63 and 0.75. The results generally agree with those from previous reviews, but water quality interventions (point-of-use water treatment) were found to be more effective than previously thought, and multiple interventions (consisting of combined water, sanitation, and hygiene measures) were not more effective than interventions with a single focus. There is some evidence of publication bias in the findings from the hygiene and water treatment interventions.

Efficacy of electrolyzed oxidizing water for the microbial safety and quality of eggs; Bialka KL et al.; During commercial processing, eggs are washed in an alkaline detergent and then rinsed with chlorine to reduce dirt, debris, and microorganism levels. The alkaline and acidic fractions of electrolyzed oxidizing (EO) water have the ability to fit into the 2-step commercial egg washing process easily if proven to be effective. Therefore, the efficacy of EO water to decontaminate Salmonella Enteritidis and Escherichia coli K12 on artificially inoculated shell eggs was investigated. For the in vitro study, eggs were soaked in alkaline EO water followed by soaking in acidic EO water at various temperatures and times. Treated eggs showed a reduction in population between > or = 0.6 to > or =2.6 log10 cfu/g of shell for S. Enteritidis and > or =0.9 and > or =2.6 log10 for E. coli K12. Log10 reductions of 1.7 and 2.0 for S. Enteritidis and E. coli K12, respectively, were observed for typical commercial detergent-sanitizer treatments, whereas log10 reductions of > or =2.1 and > or =2.3 for S. Enteritidis and E. coli K12, respectively, were achieved using the EO water treatment. For the pilot-scale study, both fractions of EO water were compared with the detergent-sanitizer treatment using E. coli K12. Log10 reductions of > or = 2.98 and > or = 2.91 were found using the EO water treatment and the detergent-sanitizer treatment, respectively. d, e, h, k. The effects of 2 treatments on egg quality were investigated. EO water and the detergent-sanitizer treatments did not significantly affect albumen height or eggshell strength; however, there were significant affects on cuticle presence. These results indicate that EO water has the potential to be used as a sanitizing agent for the egg washing process.

Polycyclic musks in the Ruhr catchment area-transport, discharges of waste water, and transformations of HHCB, AHTN and HHCB-lactone; Bester K; The polycyclic musk fragrance compounds HHCB (1,3,4,6,7,8-hexahydro-4,6,6,7,8,8-hexamethylcyclopenta-(g)-2-benzopyran; trade name, e.g. galaxolide{registered sign}) and AHTN (7-acetyl-1,1,3,4,4,6-hexamethyl-1,2,3,4-tetrahydronaphthalene, trade name, e.g. tonalide{registered sign}) and the transformation product of HHCB (HHCB-lactone) were analysed in surface water samples and sewage treatment plants (STP) effluents in the Ruhr megalopolis. The STPs were the dominant source for these pollutants. In the part of the river where the drinking water is extracted from the river, about 60 ng L(-1) HHCB, 10 ng L(-1) AHTN and 20-30 ng L(-1) HHCB-lactone were found as typical riverine concentrations, while none of the compounds were detected near the spring of the river. On the other hand sewage treatment plant effluents exhibited concentrations up to 600 ng L(-1). The STP's effluent resulted in elevated concentrations in some parts of the river and in the lakes into which they discharge. As some of the plants emit HHCB-lactone with a significantly changed enantiomeric pattern, biotransformation of HHCB to HHCB-lactone occurs in some waste water treatment plants operating with activated sludge. In those parts of the river where no relevant discharges of waste water or fresh water takes place neither the concentration nor the pattern changes significantly. This holds true especially for the HHCB versus HHCB-lactone ratios which indicates degradaticn less than 15% of the HHCB inventory in the river Ruhr itself. In other rivers, such as the Rhine, higher levels of HHCB-lactone in comparison to HHCB were detected (ratio 1 {ratio} 1).

Present and future control of cryptosporidiosis in humans and animals; Jenkins MC; Although water treatment processes to remove Cryptosporidium are improving and detection methods for identifying the parasite in water are becoming more sensitive, outbreaks of cryptosporidiosis continue in the human population. Animals, especially dairy calves, often become infected as C. parvum oocysts are present in high numbers, remain viable for long periods of time after excretion and no disinfectants, prophylactic or therapeutic reagents exist. Vaccination against C. parvum is being attempted in ruminants for the purpose of generating hyperimmune colostrum containing antibodies that may be effective in passive immunotherapy against cryptosporidiosis in the young. A number of recombinant C. parvum surface or internal antigens have been expressed by DNA-cloning technology. Immune colostrum specific for several recombinant C. parvum proteins have demonstrated efficacy in murine and ruminant models against cryptosporidiosis. The P23 and CP15 antigens appear to be the most promising candidates for vaccine development. Recent studies have demonstrated efficacy of the drug nitazoxanide against C. parvum infection in humans. In the near future, control of this parasitic disease in humans and animals will rely on a combination of passive immunotherapy and selective drug treatment.

3

Identification of a Third Sulfate Activation System in Sinorhizobium

Sinorhizobium sp. strain BR816 possesses two *nodPQ* copies, providing activated sulfate (3'-phosphoadenosine-5'-phosphosulfate [PAPS]) needed for the biosynthesis of sulfated Nod factors. It was previously shown that the Nod factors synthesized by a *nodPQ* double mutant are not structurally different from those of the wild-type strain. In this study, we describe the characterization of a third sulfate activation locus. Two open reading frames were fully characterized and displayed the highest similarity with the *Sinorhizobium meliloti* housekeeping ATP sulfurylase subunits, encoded by the *cysDN* genes. The growth characteristics as well as the levels of Nod factor sulfation of a *cysD* mutant (FAJ1600) and a *nodP1 nodQ2 cysD* triple mutant (FAJ1604) were determined. FAJ1600 shows a prolonged lag phase only with inorganic sulfate as the sole sulfur source, compared to the wild-type parent. On the other hand, FAJ1604 requires cysteine for growth and produces sulfate-free Nod factors.

Apigenin-induced *nod* gene expression for Nod factor synthesis does not influence the growth characteristics of any of the strains studied in the presence of different sulfur sources. In this way, it could be demonstrated that the "household" CysDN sulfate activation complex of *Sinorhizobium* sp. strain BR816 can additionally ensure Nod factor sulfation, whereas the symbiotic PAPS pool, generated by the *nodPQ* sulfate activation loci, can be engaged for sulfation of amino acids. Finally, our results show that rhizobial growth defects are likely the reason for a decreased nitrogen fixation capacity of bean plants inoculated with *cysD* mutant strains, which can be restored by adding methionine to the plant nutrient solution.

INTRODUCTION

Sulfur is a macronutrient that is required by all organisms. It forms constituents of proteins, lipids, carbohydrates, electron carriers, and numerous cellular metabolites. Sulfate is the most abundant source of utilizable sulfur in the aerobic biosphere. The sulfate assimilation complex, required for the formation of the sulfur-containing amino acid cysteine, has been the subject of intensive study in *Escherichia coli*. Cysteine is the central precursor of all organic molecules containing reduced sulfur, ranging from the amino acid methionine to peptides, proteins, vitamins, cofactors such as *S*-adenosylmethionine, and hormones.

Like all inorganic nutrients, sulfate is transported into cells by highly specific membrane transport systems. Sulfate assimilation requires its prior activation to adenylate compounds via a pathway that seems to be similar in all organisms. The activation is achieved by the ATP sulfurylase-catalyzed reaction of sulfate with ATP to give adenosine 5'-phosphosulfate (APS), coupled with GTP hydrolysis. Subsequently, APS is phosphorylated by an APS kinase to produce 3'-phosphoadenosine-5'-phosphosulfate (PAPS). In *E. coli*, ATP sulfurylase is encoded by *cysD* and *cysN*, whereas the APS kinase is encoded by *cysC*. PAPS is then enzymatically reduced by the *cysH*-encoded PAPS reductase (also known as PAPS sulfotransferase) to sulfite, which enters the cysteine biosynthetic pathway.

PAPS also serves directly as a sulfate donor for the formation of sulfated compounds. For example, *Rhizobium*-legume symbiotic interactions are mediated by a host-specific bacterial signaling molecule (the Nod factor), which can be sulfated. In general, rhizobial species that produce sulfated Nod factors possess at least two sulfate activation systems (6, 12, 24, 25, 40). The three genes that are indispensable for Nod factor sulfation, *nodP, nodQ,* and *nodH,* were first isolated from *Sinorhizobium meliloti*. Together, *nodP* and *nodQ* encode both ATP sulfurylase and APS kinase activities (45, 47), whereas the *nodH* gene product, a sulfotransferase, directly transfers the activated sulfate moiety to the Nod factor backbone (8, 44). NodP is homologous to *E. coli* CysD, while the amino- and carboxy-terminal domains of NodQ are homologous to *E. coli* CysN and CysC, respectively. In a recent study, it was reported that the specificity of phytopathogen-host interactions also can be controlled by a sulfated avirulence effector molecule, which is yet to be identified (48). The rice pathogen *Xanthomonas oryzae* pv. oryzae RaxP and RaxQ proteins are responsible for the synthesis of an activated form of sulfate and are similar to the NodP and NodQ host specificity proteins of the bacterial symbiont *S. meliloti*.

In *S. meliloti*, two copies of the *nodPQ* operon are present. Both copies are involved in Nod factor sulfation but are not necessary for cysteine biosynthesis. Recently, in *S. meliloti* and in *Rhizobium tropici* CFN299, homologues of the *cysDN* (ATP sulfurylase) and *cysH* (APS reductase) genes were isolated, but no homologue of the *E. coli cysC* gene (APS kinase) could be identified (1, 23). Consequently, it was demonstrated that in *S. meliloti*, APS rather than PAPS is reduced for sulfite production during cysteine biosynthesis (1). Other members of the *Rhizobiaceae*, differing in their ability to incorporate sulfate in either a Nod factor or lipopolysaccharide, also preferentially reduce APS instead of PAPS for cysteine biosynthesis. This implies that APS reduction is not necessarily correlated with the presence of PAPS-dependent sulfurylation reactions for symbiosis, which is the case when functional *nodPQ* genes are present (1). Recently, Kopriva et al. (20) have described a phylogenetic classification of APS and PAPS reductase amino acid sequences (both annotated as CysH) from different organisms. The resulting sequence-based prediction of the substrate specificities of these enzymes was confirmed by Williams et al. (58), using genetic complementation experiments.

Sinorhizobium sp. strain BR816 (formerly *Rhizobium* sp. strain BR816) synthesizes Nod factors that are fully sulfated at the reducing terminal residue (50), as is the case for the narrow-host-range *S. meliloti* (26). The sulfate decoration on the Nod factors secreted by *S. meliloti* is essential for nodulation of alfalfa (40). Except for *S. meliloti*, it is still unclear whether rhizobia producing sulfated Nod factors use only the *nodPQ*-dependent PAPS pool as a source of activated sulfate for Nod factor sulfation, the housekeeping PAPS pool, or both (25). Previously, Laeremans et al. (25) demonstrated that *Sinorhizobium* sp. strain BR816 possesses two *nodPQ* copies.

Although both copies are functional, as demonstrated by genetic complementation of an *R. tropici nodP* mutant, the double mutants did not show any detectable changes in the amount of sulfated Nod factors produced by this strain (25). It was suggested that in *Sinorhizobium* sp. strain BR816, in contrast to *S. meliloti*, a housekeeping locus as a third PAPS-producing locus could be involved in the sulfation of the Nod factors.

We have isolated the *cysDN* homologues of *Sinorhizobium* sp. strain BR816 and studied the role of this third PAPS-producing locus in relation to Nod factor synthesis. In addition, we were interested to know how the various forms of activated sulfate may be partitioned into the pathways for amino acid biosynthesis and sulfation or methylation of Nod factors and other compounds important during symbiosis. Furthermore, based on the analysis of the phylogenetic relationship among rhizobial ATP sulfurylases, we speculate on the possible origin and functionality of genes for sulfate activation.

MATERIALS AND METHODS

Bacterial Strains and Growth Conditions

The bacterial strains and plasmids used in this study are listed in Table 1. *E. coli* strains were maintained on Luria-Bertani agar at 37°C and grown in Luria-Bertani broth (32). Rhizobial strains were maintained on yeast extract-mannitol medium (55) or on tryptone-yeast medium with added $CaCl_2$ (3) at 30°C. Antibiotics were added to the medium as needed at the following concentrations (micrograms per milliliter): ampicillin, 100; spectinomycin, 50; kanamycin, 50; and nalidixic acid, 31. Tetracycline was added to a final concentration of 1 µg/ml (for *Sinorhizobium* sp. strain BR816) or 10 µg/ml (for *E. coli*). Triparental conjugations and site-directed mutagenesis were done as previously described (31).

Nucleic acid manipulations and analysis

Isolation and cloning of plasmid DNA was performed as described previously (2, 42). Total genomic DNA of *Sinorhizobium* sp. strain BR816 was isolated by using a genomic DNA isolation kit (Gentra Systems) according to the manufacturer's instructions. DNA fragments were recovered from agarose gels by using the Nucleotrap kit (Macherey-Nagel). Southern blotting and hybridizations were carried out as previously described (25). Sequencing of DNA fragments cloned in the pUC18-pUC19 vectors was performed on an automated ALF sequencer with fluorescein-labeled universal and synthetic oligonucleotide primers (Amersham Pharmacia Biotech, Uppsala, Sweden). Database searches for similarity were performed with the BLAST software (National Center for Biotechnology Information, National Institutes of Health).

PCR was performed with *Taq* DNA polymerase (Boehringer, Mannheim, Germany) according to the manufacturer's protocol. For sequencing, the high-fidelity Platinum *Pfx* DNA polymerase (GIBCO-BRL, Life Technologies) was used according to the manufacturer's protocol.

To construct a genomic minilibrary, total genomic DNA from *Sinorhizobium* sp. strain BR816 was digested with *Eco*RI. DNA fragments ranging between 4 and 6 kb were recovered and ligated into the pUC19 cloning vector. Eight hundred Ap^r white colonies were picked up. Plasmid DNA was purified from 15 pools consisting of approximately 50 colonies, and efficient insertion of fragments of the desired size was confirmed. A 450-bp PCR fragment containing an internal part of *cysD* was used as a probe to screen the library.

Phylogenetic analysis of CysD homologues

The amino acid sequences of 19 CysD-like proteins, truncated to the same size as the shortest sequence (position 3 to 299 from the *S. meliloti* NodP1 sequence [gi14523565]) were aligned by using the ClustalW program (http://searchlauncher.bcm.tmc.edu/multi-align/multi-align.html). The construction of neighbor-joining trees (41) and bootstrap analysis of 1,000 resamples were performed by using the Treecon for Windows (1.3b) software package (53). In estimating evolutionary distances between amino acid sequences, we used the Poisson correction. Insertions and deletions were not taken into account. For constructing trees by the parsimony method, the PROTPARS program in the PHYLIP package was used (10). Again, bootstrap analysis of 1,000 resamples was performed.

Growth tests

Growth tests of *Sinorhizobium* sp. strain BR816 in sulfate-free liquid medium were carried out in acid minimal salts (AMS) medium (36) containing 1 mM $CaCl_2$ with sulfate salts replaced by equimolar amounts of alternative salts ($MgCl_2$, $ZnCl_2$, $MnCl_2$, and $CuCl_2$). Ammonium chloride (10 mM) and mannitol (10 mM) were used as nitrogen and carbon sources, respectively. Sulfur compounds (sodium sulfate, sodium sulfite, L-cysteine, and L-methionine) were filter sterilized and added to the autoclaved medium at a concentration of 25 µM. When appropriate, cell cultures were induced with 500 nM apigenin. Cells of the strains tested were grown overnight in tryptone-yeast medium, washed twice in sulfate-free AMS medium, brought to an optical density of 0.4 (measured at 600 nm with a Perkin-Elmer lambda 2 spectrometer), and diluted 6,000-fold in sulfate-free AMS medium with the appropriate concentrations of filter-sterilized antibiotics, apigenin, and sulfur compounds.

Bacteria were grown in microtiter plates (final volume, 300 µl) over a 4-day period, and cell growth was monitored automatically by measuring the optical density at 600 nm in BioscreenC (Labsystems) every 30 min. For each time point, the average optical density was calculated from five independent measurements.

Insertion mutagenesis

A *Sinorhizobium* sp. strain BR816 *cysD* single mutant and *nodP1 nodQ2 cysD* triple mutant were constructed as follows. To obtain the *cysD* single mutant, the 1.6-kb *Sma*I fragment of pBRE4.8 was ligated into the *Sma*I site of pJQ200uc1. This vector allows positive selection of double homologous recombinants on sucrose (10%)-containing medium due to the presence of the *Bacillus subtilis sacB* gene.

The resulting plasmid was digested with *Bam*HI and then blunt-end ligated to the *Sma*I fragment containing the Tc^r cassette from pHP45-Tc. This plasmid was conjugated to *Sinorhizobium* sp. strain BR816. Correct insertion of the Tc^r interposon was verified by Southern hybridization with the *cysD* gene and the Tc^r cassette as probes. In this way, the same construct was introduced in CFNE205 (*nodP1*), CFNE206 (*nodQ2*), CFNE207 (*nodP1 nodP2*), and CFNE208 (*nodP1 nodQ2*) (Table 1; Fig. 1). A *cysD* single mutant (FAJ1600) and a *nodP1 nodQ2 cysD* triple mutant (FAJ1604) were obtained and retained for further analysis.

Radioactive labeling of Nod metabolites and thin-layer chromatography (TLC) analysis

Nod factors were labeled by using the isotopes [^{14}C]acetate and [^{35}S]sulfate according to a slightly modified version of the protocol of Mergaert et al. , as previously described. For this experiment, Nod factors were purified from cells grown in sulfate-free AMS minimal medium supplemented with L-cysteine, as described for the growth tests.

Plant nodulation assay

Seeds of *Phaseolus vulgaris* cv. BAT477 were surface sterilized and germinated as described previously (56). Bean seedlings were planted in 250-ml flasks containing a nitrogen-free Snoeck medium agar slant (C. Snoeck, J. Vanderleyden, and E. Schrevens, submitted for publication) with KH_2PO_4 (7.49 mM), K_2SO_4 (0.43 mM), $CaCl_2$ (2.65 mM), $MgCl_2$ (1.75 mM), $MgSO_4$ (1.2 μM), FeNaEDTA (50.8 μM), $MnSO_4$ (35.2 μM), $CuSO_4$ (0.5 μM), $ZnSO_4$ (1.5 μM), H_3BO_3 (25 μM), and $(NH_4)_6Mo_7O_{24}$ (0.07 μM), with sulfate as the sole sulfur source unless otherwise stated. The seedlings were inoculated with approximately 10^6 bacteria per plant, from a diluted overnight culture that was washed twice with sulfate-free AMS medium. The plants were maintained in a growth chamber at 26°C (day) and 22°C (night) with a 12-h photoperiod. Plants were harvested after 3 weeks. Uninoculated control plants did not show any nodules or nodule-like structures. Ten plants per strain were tested in each experiment. Nitrogenase activity was determined by measuring the acetylene reduction activity of nodulated roots in closed vessels with a Hewlett-Packard 5890A gas chromatograph equipped with a PLOT fused silica column, with propane as an internal standard.

Data analysis

In all experiments, a randomized block design was used with 10 replicate blocks. Nodule number, nodule dry weight, and acetylene reduction activity were analyzed with the means and general linear model procedure (SAS Institute, Cary, N.C.). Comparison among the mean values obtained for each strain was made by Tukey's multiple-range test with a 95% confidence limit.

Nucleotide sequence accession number

Nucleotide sequence data were deposited in the GenBank database under accession number AJ505754.

RESULTS

Cloning and sequencing of a third PAPS-producing locus in Sinorhizobium sp. strain BR816

Previous work provided evidence for the presence of a putative third PAPS-producing locus in *Sinorhizobium* sp. strain BR816 on an approximately 4.8-kb *Eco*RI genomic DNA fragment (25). In order to clone this third copy of sulfate activation genes, a genomic minilibrary was constructed (see Materials and Methods), and a single positive clone, pBRE4.8, was obtained. Since the inserted genomic DNA region corresponding to the *cysD* gene was incomplete, the missing part of *cysD* was obtained by PCR with primers that were designed based on existing knowledge of the genomic organizations and DNA sequences of sulfate assimilation genes in other *Rhizobium* spp. (1, 23).

A physical map of the 4.8-kb *Eco*RI fragment and the upstream 442-bp PCR fragment was established (Fig. 1A), and the nucleotide sequence was determined. Similarity with an ATP sulfurylase encoded by the *cysD* and *cysN* genes of *S. meliloti*, *R. tropici* CFN299, and *E. coli* was found. Partial sequence similarity upstream of the *cysD* gene revealed the presence of a *cysH* homologue, encoding an APS or PAPS reductase, whereas no *cys*C homologue was found in the sequenced fragment. The same organization is found in *S. meliloti* and *R.*

tropici (1, 23). It is likely that all three open reading frames are in a single operon, since no promoter consensus sequences or transcription termination signals were found in the intergenic *cysH-cysD* sequence of BR816. A similar situation was observed in *S. meliloti*, where two transcriptional start sites were identified, both upstream of the *cysH* homologue (1). In contrast, in *E. coli*, *cysH* does not form an operon with *cysDNC* (21). The *nodP* and *nodQ* homologues have a lower percent G+C content than the *cysD* and *cysN* homologues (data not shown), as observed for the *S. meliloti* genome (13).

The *Sinorhizobium* sp. strain BR816 *cysD* and *cysN* genes encode proteins of 317 and 498 amino acids, respectively. Strong conservation of amino acid residues was found with the respective CysD and CysN proteins of *S. meliloti* (96 and 91% identity, respectively), *R. tropici* (89 and 82% identity), and *E. coli* (68 and 52% identity). CysN contains the characteristic GTP-binding motif (GxxxxGK, DxxG, and NKxD) (7) and also an ITI motif, which is conserved among elongation factors (19). In comparison to the NodQ peptides, the deduced amino acid sequence of *cysN* lacks the carboxy-terminal part that corresponds to *E. coli* CysC. Therefore, no ATP-binding or PAPS-binding motifs were found. Similar observations were made for *S. meliloti* and *R. tropici*. In summary, these data support the ATP sulfurylase activity of the putative proteins encoded by the isolated BR816 *cysDN* genes.

Phylogenetic analysis of CysD and CysN homologues

The BR816 CysD and CysN ATP sulfurylase subunits were compared through multiple-sequence alignment (ClustalW) with homologous ATP sulfurylases subunits retrieved from GenBank. The genomic organizations of the different sulfate assimilation loci of the strains selected for the phylogenetic analysis are schematically drawn in Fig. 2. Phylogenetic analysis of *cysD* and *nodP* gene products by the protein parsimony method resulted in a maximum-parsimony tree, as shown in

Fig. 3. An identical tree topology could be inferred by using the neighbor-joining method (data not shown). Similar phylogenetic relationships could be deduced after construction of a phylogenetic dendrogram of CysN and NodQ protein sequences by using either the neighbor-joining method or protein parsimony analysis (data not shown).

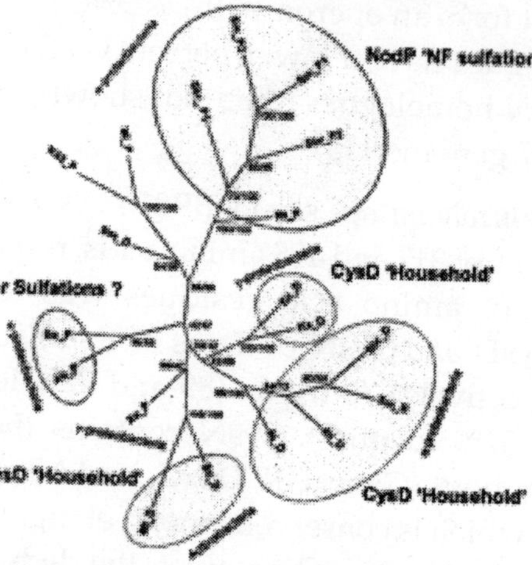

Fig. 3. Phylogenetic relationships among *cysD* gene products. The tree topology was inferred by using the protein parsimony method. Numbers represent the bootstrapping score (9) over 1,000 trials (parsimony/distance). The abbreviations of the species are as for Fig. 2.

It can be observed that the CysD and NodP ATP sulfurylase subunits of *Rhizobium* spp. producing sulfated Nod factors (Fig. 2), which have been shown to be involved in amino acid biosynthesis (1, 23) and Nod factor sulfation (24, 25, 47), respectively, cluster in two different groups (Fig. 3). The CysD protein of *Sinorhizobium* sp. strain BR816 clearly belongs to the protein cluster involved in biosynthesis of sulfur-containing amino acids, supporting its putative function.

Two other "household" clusters could be distinguished, i.e., the -*Proteobacteria* clade and the *Actinobacteria* clade. Interestingly, only one gene copy coding for a sulfate activation complex has been described, for *Mycobacterium tuberculosis* (*cysDNC*) (Fig. 2) (58). The sulfate assimilation pathway of *Mycobacterium tuberculosis* proceeds from sulfate through APS (catalyzed by CysDN), which is converted by APS reductase (CysH) in the first step toward cysteine and methionine.

APS can also be converted to PAPS, through the action of the APS kinase CysC, and serves as a substrate for sulfotransferases that produce sulfolipids, which putatively function as virulence factors (58). Similarly, APS and PAPS pools are generated through the enzymatic activity of RaxP and RaxQ in *X. oryzae* pv. oryzae and are used for both cysteine synthesis and sulfation of avirulence effector molecules (48).

The CysD-homologous proteins of some members of the *Rhizobiaceae* (among which are *Mesorhizobium loti*, producing nonsulfated Nod factors [29, 34]; *Mesorhizobium* sp. strain N33, producing sulfated Nod factors [35]; and the pathogen *Brucella melitensis*) seem to belong to another cluster.

However, these proteins are still more closely related to the NodP Nod factor sulfation cluster than to the CysD household cluster, as defined above. *Brucella melitensis* was previously shown to be genetically closely related to *Rhizobium* spp. (14). Intriguingly, the respective *Bradyrhizobium elkanii* and *Azospirillum brasilense* ATP sulfurylase subunits constitute a separate cluster (Fig. 3).

The *nodPQ* genes of *B. elkanii* are situated within a gene cluster comprising genes for symbiotic functions (*fixGHIS* and *noeE*) as well as genes involved in rhizobitoxin biosynthesis (59). Since the *B. elkanii* Nod factors are not sulfated (4, 43), these genes do not function in Nod factor biosynthesis. The recently finished genome sequencing of the p90 plasmid of *A. brasilense* sheds new light on a possible function of its *nodPQ* copy, which is located within a region carrying genes involved in polysaccharide synthesis (E. Vanbleu and J. Vanderleyden, unpublished results). It was previously shown that *A. brasilense* does not synthesize Nod factors and that deletion of the *nodPQ* copy does not lead to auxotrophy (54). Therefore, it can be speculated that this cluster encompasses proteins belonging to a novel functionality group.

Growth characteristics of Sinorhizobium sp. strain BR816 cysD mutants under free-living conditions

To investigate the biochemical role of the isolated *cysDN* genes of BR816, the BR816 *cysD* gene was mutated (see Materials and Methods). First, the *cysD* mutants were tested for cysteine auxotrophy. In addition, we were interested to know whether a *cysD* mutation could be complemented by one or both *nodP* copies of *Sinorhizobium* sp. strain BR816. Growth of the wild type and various mutants with mutations in *nodPQ* and/or *cysDN* (FAJ1600, FAJ1604, CFNE205, CFNE206, CFNE207, and CFNE208) was examined in liquid sulfate-free AMS medium supplemented with various sulfur sources (see Materials and Methods). It could be demonstrated that the BR816 *nodPQ* single or double mutants (CFNE205, CFNE206, CFNE207, and CFNE208) exhibit growth patterns similar to that of the wild-type strain in minimal medium with sulfate as the sole sulfur source (data not shown). Therefore, it can be concluded that *nodPQ* mutants are not auxotrophs. Growth of the *cysD* mutant (FAJ1600) with sulfate as the sole sulfur source was clearly affected compared to that of the wild-type strain (Fig. 4A). FAJ1600 showed a prolonged lag phase, although its generation time in exponential growth phase did not markedly differ from that of the wild type. The *nodP1 nodQ2 cysD* triple mutant (FAJ1604) was completely impaired in growth (Fig. 4A). In the presence of sulfite, cysteine, or methionine, the growth of both mutants after 60 h was nearly restored to the wild-type level (Fig. 4B to D). This indicates that the *cysDN* genes are effectively involved in the biosynthesis of sulfur-containing amino acids, more specifically in the step of the sulfate assimilatory pathway just before the reduction of activated sulfate to sulfite. From this experiment we can conclude that knocking out the three sulfate activation systems (FAJ1604) in *Sinorhizobium* sp. strain BR816 leads to cysteine auxotrophy.

Interestingly, the growth characteristics of FAJ1600 showed a course similar to that of the wild type after a certain time interval. This demonstrates that the PAPS pool generated by the NodPQ sulfate activation complex is accessible for reduction by CysH and thus is available for the biosynthesis of sulfur-containing amino acids. The growth delay of FAJ1600 might indicate that CysH of *Sinorhizobium* sp. strain BR816 preferentially shows APS reductase activity rather than PAPS reductase activity toward the formation of sulfite. Moreover, the APS reductase activity of CysH has been recently confirmed in many rhizobial species (1, 20).

One should consider that (i) the growth curves of the wild-type and mutant strains were monitored under conditions in which no Nod factors are produced (no flavonoid induction) and (ii) *nodP2*, which is localized in the nodulation region on the symbiotic plasmid, probably is *nod* box dependent and thus not expressed (49). Therefore, to investigate whether the simultaneous production of sulfated Nod factors affects growth characteristics of the *cysDN* mutant strains, similar growth tests were performed in the presence of the *nod* gene inducer apigenin and with sulfate as the sole sulfur source. In this case, similar growth courses were obtained for FAJ1600 and FAJ1604 compared to the wild type (data not shown). This implies that at least the expressed *nodPQ* copy can complement and is sufficient for growth of FAJ1600 in minimal medium with sulfate as the sole sulfur source. The use of higher concentrations of inducer did not have a significant effect on the growth curves of the strains tested.

Nod factor sulfation pattern of Sinorhizobium sp. strain BR816 cysD mutants

Since the available *nodPQ* single and double mutants of *Sinorhizobium* sp. strain BR816 (CFNE205, CFNE206, CFNE207, and CFNE208 [Table 1]) were not auxotrophic and still produced sulfated Nod factors, Laeremans et al. (25) speculated that the housekeeping *cysDN*(*C*) genes can complement mutations in genes responsible for Nod factor sulfation. In order to determine to what level the Nod factors produced by the wild-type strain and the mutant strains FAJ1600 and FAJ1604 were still sulfated, apigenin-induced cell cultures, grown in liquid sulfate-free AMS medium supplemented with cysteine, were labeled with [^{14}C]acetate or [^{35}S]sulfate, and butanol extracts of the cell cultures were analyzed by reverse-phase TLC. Separation of the BR816 Nod factors revealed the presence of apigenin-induced spots on the chromatogram, corresponding to the Nod factors of BR816 (Fig. 5). The triple mutant FAJ1604 no longer produced sulfated Nod factors, which is in clear contrast with the sulfated Nod factor pattern of both the wild-type strain and FAJ1600 (Fig. 5). These results indicate that an activated sulfate source needed for Nod factor sulfation is no longer present. It can be concluded that the *cysDN* sulfate assimilation locus does provide active sulfate for NF sulfation.

Symbiotic phenotype of cysD mutants

The *Sinorhizobium* sp. strain BR816 *cysD* mutants were tested for their ability to nodulate common bean (*P. vulgaris* cv. BAT477) and to fix nitrogen. No significant differences in the kinetics of appearance of the first nodules were observed (Fig. 6). However, FAJ1600 (*cysD*) as well as FAJ1604 (*nodP1 nodQ2 cysD*) showed a decreased nodule number per plant over time, but only for FAJ1604 was this difference significant at the 95% level (Tukey's test). Morphologically, the nodules of both mutant strains were generally smaller with apparently less leghemoglobin present (as judged by the absence of pink color).

To study the nitrogen fixation capacity of the nodulated roots, the acetylene reduction activity was measured. The acetylene reduction activity of 21-day-old nodules induced by FAJ1600 or FAJ1604 was significantly lower than that for the wild-type strain ($P < 0.05$; Tukey's test) (data not shown). When methionine was added to the plant nutrient solution, the nitrogen fixation per plant was restored to wild-type levels. Interestingly, supplementation with methionine resulted in an overall higher nitrogen fixation capacity of *P. vulgaris* cv. BAT477 inoculated with *Sinorhizobium* sp. strain BR816 (data not shown).

DISCUSSION

In this study, a third APS-producing locus of the broad-host-range strain *Sinorhizobium* sp. strain BR816 was isolated. The nucleotide sequence of this region was determined, and based on homology searches, *cysD* and *cysN* were identified. Like in *S. meliloti*, no *cysC* homologue could be isolated downstream from *cysDN*. This is an indication that, like in other rhizobia, APS rather than PAPS is reduced to sulfite for cysteine biosynthesis (1). The highest similarity was found with the *cysDN* homologues in *S. meliloti*, supporting the close phylogenetic relationship between *S. meliloti* and *Sinorhizobium* sp. strain BR816 (15). Phylogenetic analysis revealed that CysD does not cluster with NodP1 and NodP2. The two BR816 NodP proteins are closely related and could have originated from a recent gene duplication, as was proposed for the NodP proteins of *S. meliloti* (13).

Within the -*Proteobacteria* clade, two clusters of proteins are clearly functionally distinguished and were designated NodP Nod factor sulfation and CysD household. It has been demonstrated that the *nodPQ* genes are also required for sulfation of *S. meliloti* lipopolysaccharide, proving a dual functionality of members of the NodP Nod factor sulfation cluster (5, 17). A potential new NodP-like protein cluster is proposed, comprising proteins involved in sulfate activation for sulfation of compounds that are yet unknown but which could be important during symbiosis. Other closely related CysD and NodP homologous do not fit into a specific functionality group, since these proteins are involved either in sulfation of amino acids (*M. loti* and *B. melitensis*) or in sulfation of Nod factors (*Mesorhizobium* sp. strain N33).

It should be noted that within the -*Proteobacteria* clade and the *Actinobacteria* clade, only one copy of genes encoding sulfate activating enzymes is present, which seems to be involved in biosynthesis of sulfur-containing amino acids as well as sulfation of other macromolecules.

We examined the effect of a *cysD* mutation under free-living conditions in a wild-type chromosomal background and in a *nodP1 nodQ2* double mutant background.

The levels of Nod factor sulfation (Fig. 5) as well as the growth characteristics (Fig. 4) of the different mutants were determined.

In this study, we could demonstrate that the household CysDN sulfate activation locus of BR816 can additionally ensure Nod factor sulfation, whereas the symbiotic (P)APS pool, generated by the *nodPQ* sulfate activation complexes, can be engaged for sulfation of amino acids.

Figure 7 shows a model of how the various forms of activated sulfate in *Sinorhizobium* sp. strain BR816 may be partitioned into the pathways for amino acid biosynthesis and sulfation of Nod factors and other compounds that might be important during symbiosis.

The *cysDN*-dependent APS pool supplies activated sulfate that is subsequently reduced to form sulfite by the CysH APS reductase.

Sulfite is further reduced to sulfide, which is then incorporated into the cysteine and methionine biosynthesis pathway. Our data suggest that the symbiotic APS and/or PAPS pool, created by the *nodPQ*-dependent sulfate activation step, can also be used by CysH (in a less efficient manner) for the biosynthesis of sulfur-containing amino acids, when needed.

Moreover, both household and symbiotic APS pools can be mutually exchanged. In *S. meliloti*, the *nodPQ*- and *cysDN*-encoded sulfate activation systems cannot substitute for each other (46, 47).

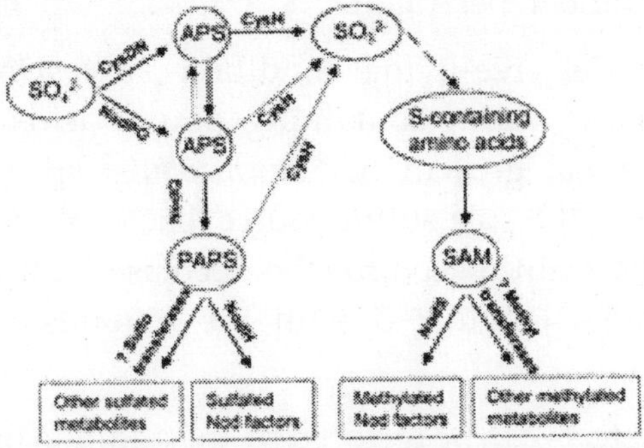

Fig. Schematic representation of the distribution of APS and PAPS for sulfation and methylation processes in *Sinorhizobium* sp. strain BR816. Dotted arrows indicate possible but less favorable enzyme activity.

Why would *Sinorhizobium* sp. strain BR816 possess three functional sulfate activation systems for Nod factor sulfation? Besides the use of activated sulfate for the biosynthesis of sulfur-containing amino acids and sulfation of Nod factors, (P)APS is needed for Nod factor methylation (37).

Introduction of the *S. meliloti nodPQ* genes into *R. tropici* resulted in a decreased rate of *R. tropici* Nod factor methylation, while all *R. tropici* Nod factor backbones were sulfated. Waelkens et al. (57) showed that methylation of Nod factors is required for nodulation of bean. In *Sinorhizobium* sp. strain BR816, the three operational sulfate-activating systems could play an important role in maintaining substitutions of bacterial determinants for symbiosis.

An *R. tropici nodPQ* mutant (producing drastically reduced amounts of sulfated Nod factors) and an *R. tropici nodH* mutant (producing nonsulfated Nod factors) still activate the signaling cascade for emergence of effective nodules on *P. vulgaris* roots (12, 24). For bean plants, the sulfate moiety of the Nod factor was shown to be involved in the efficiency of nodule formation but appears not to be essential (11, 22). The effects of the *cysD* mutant FAJ1600 and the *nodP1 nodQ2 cysD* triple mutant FAJ1604 on bean symbiosis were seen mainly in the reduction of nodule number per plant.

Since under free-living conditions, a *cysDN*-dependent biosynthesis of sulfur-containing amino acids is essential to allow optimal growth of *Sinorhizobium* sp. strain BR816 with sulfate as the sole sulfur source, bacterial growth defects are likely the main reason for the decreased nitrogen fixation of bean plants inoculated with the mutants FAJ1600 and FAJ1604.

These defects can be restored by the addition of methionine to the plant nutrient solution. We propose that at the early stages of the nodulation, the plant root exudates of the germinated seedlings provide enough sources of organic sulfur to allow bacterial growth.

However, a shortage of an organic sulfur source like methionine impairs bacterial growth inside the plant. Inoculation experiments with a *Rhizobium etli metZ* (O-succinylhomoserine sulfhydrylase for methionine biosynthesis) (51) mutant on bean plants resulted in the formation of ineffective (Nod^+ Fix^-) nodules, which suggested that root cells do not supply the inoculant bacteria with enough methionine.

The fact that supplemented methionine resulted in an overall higher nitrogen fixation capacity of *P. vulgaris* BAT477 inoculated with BR816 strains supports this hypothesis.

In contrast to our observations, an *R. etli cysG* (siroheme synthetase for cysteine biosynthesis) mutant, which is able to induce the formation of effective nodules (Nod^+ Fix^+) on the roots of common bean, seems to dispose of an organic sulfur source like cysteine or glutathione to allow growth inside the plant (52).

How can the strictly separated symbiotic and endogenous (P)APS pools in *S. meliloti* versus the complementary (P)APS pools in *Sinorhizobium* sp. strain BR816 be explained? Presumably, the *nodPQ* genes arose in ancestral rhizobial strains through duplications of the endogenous *cysDNC* genes. Later, these *nodPQ* genes evolved toward more specialized symbiotic genes, whereas the endogenous *cysC* gene, encoding the APS kinase, was apparently lost during evolution. At this stage, complementation between both PAPS pools was still possible (the case of *Sinorhizobium* sp. strain BR816). Then, the genetic separation of the two sulfate-activating systems could have further evolved into two more efficient and energy-saving separate enzymatic multienzyme complexes (the case of *S. meliloti*).

4

Resuscitation of Salmonella Enterica

Salmonella enterica serovar Typhimurium and enterohemorrhagic *Escherichia coli* were stressed by prolonged incubation in water microcosms until it was no longer possible to observe colony formation when samples were plated on nonselective medium.

Overnight incubation of samples in nutrient-rich broth medium supplemented with growth factors, however, allowed resuscitation of stressed and viable but nonculturable cells so that subsequent plating yielded observable colonies for significantly extended periods of time.

The growth factors were (i) the trihydroxamate siderophore ferrioxamine E (for *Salmonella* only), (ii) the commercially available antioxidant Oxyrase, and (iii) the heat-stable autoinducer of growth secreted by enterobacterial species in response to norepinephrine.

Analysis of water microcosms with the Bioscreen C apparatus confirmed that these supplements enhanced recovery of cells in stressed populations; enterobacterial autoinducer was the most effective, promoting resuscitation in populations that were so heavily stressed that ferrioxamine E or Oxyrase had no effect. Similar results were observed in Bioscreen analysis of bacterial populations stressed by heating. Patterns of resuscitation of *S. enterica* serovar Typhimurium *rpoS* mutants from water microcosms and heat stress were qualitatively similar, suggesting that the general stress response controlled by the ^{s} subunit of RNA polymerase plays no role in autoinducer-dependent resuscitation. Enterobacterial autoinducer also resuscitated stressed populations of *Citrobacter freundii* and *Enterobacter agglomerans*.

INTRODUCTION

Salmonella species and enterohemorrhagic *Escherichia coli* (EHEC), in particular serovar O157:H7, are important food-borne pathogens that represent an increasingly significant public health issue in industrialized countries. The problem, at least in part, is that these organisms can persist for long periods in the environment in a heavily stressed state known variously, and often contentiously, as viable but nonculturable (VNC) or not immediately culturable. These heavily stressed microorganisms show only very weak metabolic activity, often at the very limits of detection, and they lose the ability to form colonies on nonselective plating media or to grow in nonselective broth media.

Nevertheless, in the case of nonculturable populations of pathogenic bacteria in the environment (soil or water, etc.) or associated with bacteriological spoilage of human foods and animal feeds, they may still be capable of causing disease if ingested by a susceptible animal host. The important questions are how such cells can be resuscitated to aid in vitro identification of potential pathogens and what role they play in the various habitats in which they exist and in the pathogenesis of infectious disease.

We previously demonstrated that the microbial trihydroxamate siderophore ferrioxamine E was able to resuscitate stressed *Salmonella enterica* serovar Typhimurium strains in soil for periods of up to 2 years and in water microcosms for more than 4 months. This process was shown to be dependent on siderophore uptake, since neither *Salmonella* mutants deficient in ferrioxamine uptake nor strains of *E. coli* that naturally cannot use ferrioxamines as sources of iron were resuscitated from stressed states by ferrioxamine E. We proposed that resuscitation by ferrioxamine E is due to its ability to sequester intracellular iron in a form that reduces the risk of generating damaging oxygen radicals that would otherwise kill bacterial cells emerging from stressed states. Consistent with this proposal, we demonstrate here that the commercially available antioxidant Oxyrase, an oxygen radical-destroying enzyme prepared from *E. coli*, is also able to resuscitate stressed populations of *S. enterica* serovar Typhimurium and EHEC.

The aim of the work described in this paper was to assess the ability of a heat-stable autoinducer (AI) of growth, which is secreted by a number of enterobacterial species in the presence of the mammalian neuroendocrine hormone norepinephrine (NE), to resuscitate stressed bacteria.

The ability of NE and other catecholamines to promote bacterial growth in vitro in a medium designed to reflect the harsh environments infectious bacteria may encounter in vivo is well established.

In the case of many gram-negative species, such growth is accompanied by production of AI, which can itself stimulate bacterial growth under stress conditions in the absence of NE and also the production of more AI. Characterization of the activity produced by several enterobacterial species suggested a family of structurally closely related but functionally identical molecules that we refer to in the generic sense as enterobacterial AI. Here we demonstrate that enterobacterial AI is able to resuscitate heavily stressed populations of a number of strains of *S. enterica* serovar Typhimurium, EHEC, and other pathogenic species, and we propose the routine use of AI for more effective screening for enterobacterial pathogens in environmental and food samples.

MATERIALS AND METHODS

Bacterial strains

Of the species reported in our previous study of NE-induced AI, the strain that reproducibly produced the highest AI levels was one that was originally reported by the Leicester Public Health Laboratory Service to be *Hafnia alvei* on the basis of API tests. However, in subsequent independent tests with the Bactid identification system (Centers for Disease Control, Atlanta, Ga.), which in our hands is more reliable, this strain was identified as *Yersinia ruckeri*. In addition, it was positive in a pyrase test, which is characteristic of *Yersinia* spp., but negative in a highly sensitive test for *Hafnia*. For technical convenience we chose this strain as the routine source of heat-stable enterobacterial AI, and for consistency all of the experiments reported in this paper were performed with batches of AI prepared from this strain. It should be noted, however, that AI prepared from cultures of various strains of *E. coli*, *S. enterica* serovar Typhimurium, and *H. alvei* gave qualitatively similar data in all experiments in which they were used. Details of the *Salmonella* and *E. coli* strains tested for resuscitation by AI are given in Table 1. The sources of strains of other species tested are indicated elsewhere in the text.

AI preparation

SAPI medium containing 30% (vol/vol) adult bovine serum and 50 µM L-(-)-NE (Sigma Chemical Co.) as previously described was used for the preparation of enterobacterial heat-stable AI. SAPI is a variation of the standard American Petroleum Institute medium and contains 6.25 mM NH_4NO_3, 1.84 mM KH_2PO_4, 3.35 mM KCl, 1.01 mm $MgSO_4$, and 2.77 mM glucose (pH 7.5). *Y. ruckeri* was inoculated at approximately 100 to 1,000 CFU/ml in serum-SAPI medium and incubated overnight at 37°C in a humidified 5% CO_2 atmosphere. Bacteria were pelleted by centrifugation (6,000 x *g* for 15 min), and the culture supernatants were filter sterilized and stored at -20°C until required. Sterility was checked by plating samples on sheep blood agar and incubating at 37°C for 48 h under aerobic and anaerobic conditions. Sterile preparations were serially diluted in fresh sterile SAPI medium, and samples of each dilution were added at 5% (vol/vol) to serum-SAPI medium inoculated with 100 to 1,000 CFU of an indicator *E. coli* strain per ml. The dilution that promoted bacterial growth after overnight incubation at 37°C to an optical density at 620 nm of 0.4, which represents 10^8 CFU/ml, was used as a supplement to liquid growth media as described below. NE, SAPI, and bovine serum had no significant effect when used as supplements, either individually or combined.

Purified enterobacterial AI is available as Bacxell (BioNutrix LLC, Minneapolis, Minn.) (International Patent Application no. WO 98/53047 [May 1998] and United Kingdom Patent Application no. 0120120 [August 2001]).

Water microcosms

Water microcosms were set up as previously described. Briefly, 1.5-liter batches of sterile double-distilled water were inoculated with saline-washed bacterial growth from fresh tryptic soy agar (TSA) (BD Heidelberg, Heidelberg, Germany) cultures at 10^5 to 10^6 CFU/ml and stored at room temperature under normal laboratory light conditions. Samples were taken at intervals for viable count assays, growth in liquid culture, or Bioscreen C analysis as appropriate (see below). Water microcosm experiments were typically of several months' duration in the case of the gram-negative species tested, but gram-positive organisms died within a few days under these conditions.

Colony counts

Culturable counts of samples from water microcosms were measured by plating serially diluted samples in triplicate on TSA containing 0.1% (wt/vol) sodium pyruvate to minimize further selective stress by the medium. In the case of *S. enterica* serovar Typhimurium microcosms, recovered colonies were checked by subculturing onto Galle-Chrysoidin-glycerol (GCG) agar (SIFIN, Berlin, Germany) and by agglutination with omnivalent *Salmonella* serum. For *E. coli* microcosms, colonies on GCG agar and BCM *E. coli* O157:H7(+) selective agar medium (Biosynth AG, Staad, Switzerland) were confirmed serologically. Other species of the *Enterobacteriaceae*, *Pseudomonas aeruginosa*, *Aeromonas hydrophila*, and *Burkholderia cepacia* were checked by subculturing onto GCG agar. Gram-positive bacteria were characterized by typical growth on sheep blood agar.

Growth in liquid media

When colonies were no longer observed by direct viable counting of 0.1- and 0.5-ml samples, 60-ml samples of water microcosms were inoculated into 90 ml of 1.67-fold-concentrated buffered peptone water (BPW) (Oxoid) in the case of *Salmonella* strains or phosphate-buffered tryptic soy broth (p-TSB) (containing 1.5 mg of K_2HPO_4/ml) for the other bacterial species tested. Experiments to test the effects of supplementation were performed only if no growth had occurred in this mixture after overnight incubation. Ferrioxamine E (Novartis AG, Basel, Switzerland) was added at a final concentration of 50 ng/ml, Oxyrase (Oxyrase Inc., Mansfield, Ohio) was added at 0.2 U/ml as defined in the manufacturer's technical bulletin, and enterobacterial heat-stable AI (prepared as described above) was added as a 1% (vol/vol) supplement. Suspensions were incubated with rotary shaking at 37°C, except in the case of *P. aeruginosa* cultures, which were incubated at 30°C. Bacteria were inoculated onto TSA, and growth was checked biochemically, serologically, or morphologically as described above.

Heat stress

Freshly cultivated bacteria from TSA plates were suspended in phosphate-buffered saline to a density of approximately 10^8

to 10^9 CFU/ml as estimated by turbidity measurements. Viable cell counts at the beginning of the stress period were enumerated by the most-probable-number technique. Bacterial suspensions were incubated in a shaking water bath (GFL1086; Gesellschaft für Labortechnik mbH, Burgwedel, Germany) at the temperatures and for the times indicated in Table 2.

Bioscreen analysis

Three-hundred-microliter volumes of BPW or p-TSB as appropriate, with or without supplementation as described above, were applied to each of 25 wells of a 100-well microtiter plate (i.e., to provide 25 replicates of each experimental condition). Ten-microliter aliquots of stressed cell suspensions (from water microcosms or after heat stress) were inoculated into each well, and bacterial growth was monitored in a Bioscreen C apparatus (Labsystems, Helsinki, Finland). Optical densities at 620 nm values were measured at 20-min intervals (sample plates were shaken for 10 s before each measurement) over a period of 24 to 48 h at 37°C, or at 30°C in the case of *P. aeruginosa*. The measurements were monitored by the computer program DOS of the Bioscreen C apparatus and transferred to a Windows 98 platform for graphical presentation. In the graphs shown, each line represents growth in a single well; however, for reasons having to do with the operating system of the computer that records the data, one line in each panel is markedly thicker than all of the others. Bacteria from two wells of each incubation regimen in which resuscitation had occurred were streaked onto nonselective agar, and their identities confirmed biochemically, serologically, or morphologically as described above.

Reproducibility

Since the effects of stress are essentially random at the level of the individual cell, each experiment generated a unique population that responded to resuscitation quantitatively differently from other stressed populations even if apparently identical stress conditions were used. Thus, all data shown in this paper are representative of at least three replicate experiments in which qualitatively similar results were obtained.

RESULTS

Reactivation and resuscitation of stressed S. enterica serovar Typhimurium from water microcosms

Water microcosms were inoculated at 10^5 to 10^6 CFU/ml and sampled at intervals to monitor the decline in numbers of viable bacteria recoverable by direct plating on TSA. Results were variable from experiment to experiment for any individual strain and between strains, presumably reflecting natural variation in bacterial populations undergoing stress and damage, as noted in Material and Methods. In the case of wild-type *S. enterica* serovar Typhimurium strains, viable cells were no longer detectable by plating after periods ranging from about 1 month to almost 4 months.

Typically, *S. enterica* serovar Typhimurium produced small irregular colonies on the plating medium towards the end of this period, but subculture of these colonies on TSA gave normal-sized smooth colonies after one or two passages. Stressed cells that were no longer detectable by direct plating could be reactivated by overnight incubation of microcosm samples at 37°C in BPW, and were therefore detectable for an additional period of 11 to 18 days. We distinguish experimentally between microcosms containing stressed cells that can be reactivated by a period of incubation in nutrient-rich broth medium, such as BPW or p-TSB, and those in which stressed cells can be resuscitated only by supplementation with particular growth factors. As we reported previously, addition of ferrioxamine E to BPW resulted in prolonged periods of resuscitation of *S. enterica* serovar Typhimurium from highly stressed populations beyond the time when incubation in BPW alone no longer resulted in reactivation. Resuscitation by ferrioxamine E is dependent on siderophore uptake, since recovery of the *fhuC* mutant TA2700, which is unable to transport ferrioxamines, was not enhanced by the presence of ferrioxamine E. Furthermore, for all strains of *Salmonella* that we have tested, including TA2700, supplementation of BPW with enterobacterial AI resulted in resuscitation of cells in stressed populations for longer periods of time than did ferrioxamine E supplementation.

Reactivation and resuscitation of stressed EHEC from water microcosms

The viability of EHEC strains of serotype O157:H7 in water microcosms also declined with time, so that at about 3 months no growth was detectable by direct plating. Here again, small irregular colonies were observed on the plating medium just before this time. Incubation of microcosm samples in nutrient-rich p-TSB enabled reactivation of stressed cells for additional periods of up to a week. *E. coli* cannot use ferrioxamines to acquire iron, and so stressed EHEC cells cannot be resuscitated by supplementation with ferrioxamine E as in the case of *Salmonella*.

However, supplementation of p-TSB with the biological antioxidant Oxyrase resuscitated highly stressed EHEC beyond the time that incubation in unsupplemented p-TSB no longer gave visible growth. Moreover, supplementation with enterobacterial AI enabled resuscitation of stressed *E. coli* O157:H7 cells for significantly longer periods, in one experiment for as much as an additional month. In an experiment with EHEC O8:H$^-$ strain 97-04951, growth was detectable following AI supplementation of p-TSB for up to 455 days, compared with only 88 days in the absence of supplementation (data not shown). In all of these experiments, EHEC bacteria recovered after resuscitation in AI-supplemented p-TSB still produced Shiga toxins (data not shown).

Bioscreen analysis of resuscitation of stressed bacteria from water microcosms

Because the development of stress in water microcosms is an essentially random process, prolonged incubation presumably leads to progressive changes in the proportions of (i) stressed cells that can be reactivated in liquid culture, (ii) VNC cells that require a growth supplement (ferrioxamine E, Oxyrase, or enterobacterial AI) for resuscitation, and (iii) dead cells. In fact, it is not possible to predict from experiment to experiment whether a particular population sample contains any bacteria capable of reactivation or resuscitation. To analyze stressed populations in greater detail, we used the Bioscreen C apparatus to monitor the effects of various supplements on 25 replicate subcultures of a population simultaneously in real time. In an attempt to standardize the experimental approach, we tested water microcosms at the time that direct plating just failed to detect colony-forming bacteria, but we used quantitative data from the last day on which colonies were detectable to adjust the notional inoculum size (usually by concentrating samples of microcosm) to 0.3 viable bacterium per well of the Bioscreen apparatus.

The effects of supplementation on cells of *S. enterica* serovar Typhimurium ATCC 14028 stressed to the extent that growth was observed in 6 of the 25 wells containing unsupplemented BPW. In this particular experiment, only a few cells in the microcosm were resuscitated by ferrioxamine E (8 of 25 wells) or Oxyrase (3 of 25 wells), although it should be noted that the lag phases of some of these subcultures were significantly shorter than those in unsupplemented subcultures (11 h, compared with 14 h or more in the absence of supplementation).

However, 13 of the 25 replicate samples supplemented with enterobacterial AI showed resuscitation, some, remarkably, with lag phases of as short as 4 to 7 h. Note that variation between lines representing a particular treatment presumably indicates natural, random variation within the population of stressed bacteria, with longer lag phases and lower exponential growth rates presumably being due to greater levels of stress or damage in the cells of each inoculum.

Moreover, variation between experiments presumably also represents natural, random variation in the development of stress in different microcosms even when these were set up identically. Increasing stress presumably leads to conditions in which dead cells predominate in the population. In an experiment with *S. enterica* serovar Typhimurium strain ATCC 14028-1s, the level of stress was such that neither the wells with unsupplemented BPW nor the wells supplemented with ferrioxamine E or Oxyrase showed any evidence of reactivation or resuscitation. Enterobacterial AI, however, stimulated growth in 10 of the 25 wells with lag phases ranging from 5 to 19 h. The use of combinations of ferrioxamine E or Oxyrase with AI did not significantly increase the level of resuscitation observed with AI alone.

To determine whether RpoS plays a role in resuscitation, we tested microcosms of *S. enterica* serovar Typhimurium strains carrying a mutation in the *rpoS* gene. These strains lost the ability to form colonies more rapidly that wild-type strains (in 2 weeks as opposed to several months), but resuscitation patterns with ferrioxamine E, Oxyrase, or enterobacterial AI were remarkably similar to those observed for wild-type strains. Table 2 shows the effects of supplementation on stressed cells of *rpoS* mutants MAE40 and SF1005 in water microcosms; AI resulted in growth in more wells and with markedly shorter lag phases than ferrioxamine E or Oxyrase.

Bioscreen analysis of resuscitation of heat-stressed bacteria

While water microcosms are good models for the stresses that pathogenic bacteria may encounter in the external environment, survival from heat stress is an important factor in the etiology of food-borne illness due to faulty food processing and poor kitchen hygiene. An experiment in which incubation of *E. coli* strain NCTC10418 at 53°C for 50 min resulted in a population from which 7 of 25 subculture wells contained cells that could be reactivated by incubation in p-TSB. In this case, supplementation with Oxyrase or enterobacterial AI promoted resuscitation in 2 and 11 wells, respectively.

In an experiment with another EHEC strain, 97-04281, heated at 53°C for 35 min, the level of stress was so great that no growth was observed in any of the unsupplemented p-TSB wells or in any of the wells supplemented with Oxyrase. Supplementation with AI, however, allowed resuscitation in five wells from this highly heat-stressed population. Similar data were obtained with a heat-stressed population of *S. enterica* serovar Typhimurium strain UMR1 in which more prolonged periods of heating were used; in this case enterobacterial AI resulted in resuscitation of stressed cells in 11 of the 25 wells.

AI also resuscitated heat-stressed cells of the *Salmonella rpoS* mutant MAE40 with a pattern similar to those observed for resuscitation of water microcosms. Lag phases of just 9 h were observed in some wells, compared with 12 h in unsupplemented cultures. Enterobacterial AI also effectively resuscitated heat-stressed populations of wild-type *Citrobacter freundii* strains 98-01367 (from the collection of the Robert Koch Institute) and ATCC 8090 and of *Enterobacter agglomerans* ATCC 13020, but it had no measurable effects (in terms either of numbers of wells or of lag phases of resuscitated cultures) on *Proteus vulgaris* 718/96, *Proteus mirabilis* NM12, *Providencia rettgeri* NM19, *Providencia stuartii* 20137, and *Morganella morganii* SBK3 (all from the Robert Koch Institute collection); on *A. hydrophila* ATCC 7966; or on *P. aeruginosa* DSM27853 and *B. cepacia* DSM7288 (both from the Deutsche Sammlung für Mikroorganismen und Zellkulturen, Braunschweig, Germany) (data not shown). Moreover, AI was not able to resuscitate the gram-positive organisms *Staphylococcus aureus* NCTC6571 and *Staphylococcus epidermidis* CCM2124 (from the Culture Collection of Microorganisms, Brno, Czech Republic), *Enterococcus faecalis* ATCC 29212, and *Listeria monocytogenes* NCTC7073.

DISCUSSION

It is axiomatic that only a tiny fraction of the bacterial species inhabiting the biosphere have so far been discovered. Moreover, among the minority of bacteria that have been discovered, more than 90% are as yet nonculturable and can be detected only by molecular techniques based on probes for 16S and 23S rRNAs or on determination of mRNA, either by reverse transcriptase PCR or by fluorescence techniques such as in situ hybridization, microradiography, epifluorescence microscopy, and flow cytometry. Even among the minority of species that are normally considered to be readily culturable, environmental stresses of various kinds drive populations of microorganisms towards a state in which increasing proportions cease to be culturable by the use of known culture media and conditions. In the case of pathogenic species of bacteria, this so-called VNC state is a potentially dangerous public health problem, particularly because stressed cells are apparently more virulent that well-fed bacteria. Bacteria may be damaged by a wide variety of stress conditions, including nutrient starvation, oxygen radicals, heat, freezing temperatures, changes in pH, near-UV radiation, and osmotic pressure. Adaptive networks have evolved in bacteria to overcome the challenges of rapidly changing environments and to permit survival under conditions of stress. The important practical question is how to monitor highly stressed cells in the environment and the food chain.

In this paper we report the ability of ferrioxamine E, Oxyrase, and enterobacterial AI to resuscitate bacterial cells in populations stressed by prolonged incubation in water or by heat treatment. Two independent approaches were used to assess resuscitation. In one case, microcosm samples taken at intervals were cultivated in nutrient-rich medium with or without the supplements. This is an "all-or-nothing" approach that simply detects the presence of any recoverable cells in the population (theoretically a single such cell would be sufficient), but it is useful as an indicator of the presence of cells that may, given the right set of environmental parameters, go on to cause disease in susceptible individuals, human or animal, that ingest them. The second approach was to use the Bioscreen to analyze the effects of supplementation simultaneously in 25 subpopulations of a stressed population. This method enables comparisons to be made between populations at various arbitrarily defined levels of stress, from relatively mild (in which reactivation occurred in most or all of the subpopulations even in the absence of supplements) to severe (in which some subpopulations contained VNC cells but most contained only dead cells).

The important practical point to note is that despite differences in readout, the data obtained from the two approaches were entirely consistent. Thus, we confirmed our previous observation that ferrioxamine E can resuscitate wild-type *S. enterica* serovar Typhimurium from water microcosms and extended it by demonstrating a similar effect on heat-stressed populations. Since *Salmonella* strains of subspecies I, II, and IIIb, which include more than 98% of all clinical isolates, possess the high-affinity ferrioxamine uptake and utilization system, supplementation with ferrioxamine E can be considered to be semiselective for detection of *Salmonella* spp. in stressed populations.

Ferrioxamine B functions similarly but slightly less effectively than ferrioxamine E (data not shown). Oxyrase was similarly able to resuscitate stressed *S. enterica* serovar Typhimurium strains and was also effective on EHEC populations both from water microcosms and after heat stress. Oxyrase functions by destroying oxygen radicals in growing cells, and so these observations are consistent with our proposal that ferrioxamine E acts by preventing damaging oxygen radicals from killing recovering cells. However, while Oxyrase shows promise as an effective growth supplement for rapid recovery of stressed cells from various contaminated materials, unlike ferrioxamine E it would be expected to work nonselectively on both pathogenic and nonpathogenic bacteria in a population and so would not be useful in situations where the detection of particular pathogens is required.

More effective than either ferrioxamine E or Oxyrase at resuscitating bacteria from heavily stressed populations was the enterobacterial AI produced in serum-SAPI medium in the presence of NE.

This medium was designed to mimic the kind of stressful environment an infectious microorganism might encounter in vivo, and indeed most bacterial species fail to thrive in it unless NE or other catecholamines are present to facilitate iron removal and uptake from serum transferrin.

Enterobacterial AI also promotes bacterial growth in serum-SAPI; the mechanism is as yet unknown but is undoubtedly different from that of catecholamines. Thus, growth of clinical isolates of 17 gram-negative and 6 gram-positive bacterial species in serum-SAPI medium was stimulated by supplementation with AI preparations from several enterobacterial species.

Note that in these previous experiments, serum-SAPI medium was inoculated with bacteria from fresh, nonstressed starting cultures. The present study, by contrast, suggests that the ability of enterobacterial AI to resuscitate heavily stressed populations is restricted to *S. enterica* serovar Typhimurium, EHEC, *C. freundii*, and *E. agglomerans*. AI had no measurable effect on individual representative strains of eight other gram-negative species (*P. vulgaris, P. mirabilis, P. rettgeri, P. stuartii, M. morganii, A. hydrophila, P. aeruginosa*, and *B. cepacia*) and four gram-positive species (*S. aureus, S. epidermidis, E. faecalis*, and *L. monocytogenes*) that we tested.

The mechanism of action of AI is not yet known. One possibility is that it functions as a quorum sensor for stressed bacteria; the structure of AI is still under investigation, but it is known not to be a homoserine lactone, nor does it have properties similar to those of the recently characterized autoinducer-2 molecule. Another possibility is that AI interacts with components of bacterial global stress responses. A number of bacterial virulence factors are known to be activated during the transition from exponential- to stationary-phase growth in response to induction of *rpoS*. Moreover, bacteria in stationary phase are more thermotolerant, more resistant to oxidative stress and acidic conditions, and better equipped to survive both osmotic stress and starvation due to differential production of products such as catalase, glycogen, and heat shock and cold shock proteins. The subunit of RNA polymerase (RpoS) is the master regulator of more than 35 genes involved in the general stress response in *S. enterica* serovar Typhimurium, *E. coli*, and other enteric bacteria. Levels of RpoS are low in rapidly growing cells that have not been exposed to any particular stress but are induced in response to a variety of diverse environmental stresses.

Inactivation of RpoS renders cells much more sensitive to a variety of stresses, as seen by the more rapid decline in cell counts in water microcosms compared with wild-type strains of *S. enterica* serovar Typhimurium. The qualitative pattern of resuscitation of *S. enterica* serovar Typhimurium *rpoS* mutants by ferrioxamines, Oxyrase, or AI, however, was essentially the same as with wild-type strains, indicating that the mechanism(s) of resuscitation is independent of RpoS. Similarly, RNA polymerase subunit RpoB, which is not involved in the stress response, appears not to be involved in recovery from the VNC state, since an *S. enterica* serovar Typhimurium *rpoB* mutant showed the same pattern of resuscitation as wild-type strains (data not shown). In order to learn more about the mechanism of recovery from stress, we are currently performing a detailed analysis of reproductive, deenergized, depolarized, and dead cells in stressed and resuscitated populations by using various fluorescent dyes measured by flow cytometry.

Meanwhile, from the practical point of view, we define the process of resuscitation as the ability of VNC cells to grow on nonselective agar only following incubation in nutrient-rich broth medium containing supplements such as ferrioxamine E (if appropriate), Oxyrase, or enterobacterial AI. Our data demonstrate that AI is able to promote significant resuscitation of populations that were so heavily stressed that the other supplements were ineffective. Moreover, EHEC cells resuscitated by AI still retained the capacity to produce Shiga toxins (in the case of one serotype O8:H$^-$ strain, after a period of 15 months in a water microcosm). We propose that enterobacterial AI has important benefits, compared with unsupplemented preenrichment media, for improved semiselective isolation of pathogenic *Salmonella* spp. and *E. coli* O157:H7 (and other EHEC) from contaminated environments and foods.

5

Contribution of dps to Acid Stress Tolerance and Oxidative Stress Tolerance

An *Escherichia coli* O157:H7 *dps::nptI* mutant (FRIK 47991) was generated, and its survival was compared to that of the parent in HCl (synthetic gastric fluid, pH 1.8) and hydrogen peroxide (15 mM) challenges. The survival of the mutant in log phase (5-h culture) was significantly impaired (4-$\log_{10}$-CFU/ml reduction) compared to that of the parent strain (ca. 1.0-$\log_{10}$-CFU/ml reduction) after a standard 3-h acid challenge. Early-stationary-phase cells (12-h culture) of the mutant decreased by ca. 4 $\log_{10}$ CFU/ml while the parent strain decreased by approximately 2 $\log_{10}$ CFU/ml.

No significant differences in the survival of late-stationary-phase cells (24-h culture) between the parent strain and the mutant were observed, although numbers of the parent strain declined less in the initial 1 h of acid challenge.

FRIK 47991 was more sensitive to hydrogen peroxide challenge than was the parent strain, although survival improved in stationary phase. Complementation of the mutant with a functional *dps* gene restored acid and hydrogen peroxide tolerance to levels equal to or greater than those exhibited by the parent strain. These results demonstrate that decreases in survival were from the absence of Dps or a protein regulated by Dps. The results from this study establish that Dps contributes to acid tolerance in *E. coli* O157:H7 and confirm the importance of Dps in oxidative stress protection.

INTRODUCTION

Escherichia coli O157:H7 causes hemorrhagic colitis in humans and in some cases may incite hemolytic-uremic syndrome (23, 24). Data from epidemiological investigations indicate that as few as 10 to 100 cells of *E. coli* O157:H7 per g of raw ground beef are sufficient to cause illness (1, 4, 14). Additionally, person-to-person transmission has occurred in day care facilities, and waterborne transmission has resulted from swimming in contaminated waters. Collectively, these and other epidemiological investigations establish that this pathogen has a low infectious dose. Gordon and Small suggested that human pathogens with a low infectious dose that are transmitted by the fecal-oral route are acid tolerant because they must survive passage through the gastric barrier. The acid tolerance of serotype O157:H7 strains of *E. coli* is further supported by outbreaks involving acidic foods and laboratory studies.

Acid tolerance can be classified into three main strategies for bacteria. The first is changes in membrane composition, the second is enzymatic or physiological maintenance of internal pH, and the third is repair and/or prevention of damage caused to essential cellular components by acidic pH. Previous studies with *E. coli, Salmonella enterica* serovar Typhimurium, and *Helicobacter pylori* suggest that DNA repair pathways play a role in survival in extreme-acidity conditions such as the gastric barrier.

It has been shown previously that mutations in DNA repair mechanisms such as *recA* and *uvrB* in *H. pylori* resulted in significant decreases in tolerance for low pH, and these results establish the importance of DNA repair systems in survival in acidic conditions.

Damage to DNA can occur at many different sites depending on the reactive substance. Oxidative damage is characterized by the production of hydroxyl radicals that leads to oxidation of sugar and base moieties that can cause strand breaks in DNA. Low pH causes DNA damage primarily by removal of purine bases and to a lesser extent by production of double-stranded lesions in the DNA. Depurination results in unrepaired DNA and mismatches in repaired sequences, which ultimately can be lethal to the cell. DNA repair represents a major strategy for bacteria to remain viable following passage through extreme pH conditions such as the gastric barrier. This emphasizes the need to further characterize DNA protection and repair systems and determine their significance in acid tolerance.

Almirón et al. identified a DNA-binding protein, designated Dps (DNA-binding protein from starved cells), that is produced primarily in stationary-phase cells of *E. coli* and has been shown to be regulated by, and OxyR. Dps forms spherical dodecamers, homologous to ferritins, that sequester and protect DNA from oxidative stress, nucleases, and UV light. Dukan and Touati showed that mutations of *recA*, *recB*, and *dps* in *E. coli* rendered cells more sensitive to damage from hydroxyl radicals generated by HOCl. This suggests that not only DNA repair, but also DNA protection by *dps*, is pivotal for survival in extreme conditions. Therefore, the role of *dps* in protection from acid-mediated DNA damage (depurination) and its contribution to the acid tolerance of *E. coli* O157:H7 were ascertained. To this end, an *E. coli* O157:H7 *dps* null mutant was constructed, and its acid stress and oxidative stress tolerance was compared to that of the parent strain.

MATERIALS AND METHODS

Bacterial strains and culture media

The bacterial strains and plasmids used in this study are listed in Table 1. All strains were stored in nutrient broth

(Difco Laboratories, Detroit, Mich.) supplemented with 10% glycerol at 70°C. Growth of cultures was monitored spectrophotometrically at 600 nm using a Bioscreen Analyzer (Labsystems, Helsinki, Finland). For log-phase cultures, 5 µl of an 18-h culture was used to inoculate 5 ml of Luria-Bertani (LB) broth [with kanamycin for FRIK 47991 and kanamycin and ampicillin for 47991(pSC9915) and 47991(pSC9916)] and incubated with shaking (150 rpm) at 37°C for 5 h (A_{600} = 0.8). For stationary-phase or late-stationary-phase cultures, LB broth was inoculated as described previously and incubated with shaking (150 rpm) at 37°C for 12 or 24 h (A_{600} = 1.1), respectively. Transformants and mutants of *E. coli* were recovered on LB agar or MacConkey sorbitol agar (MSA). Antibiotics (Sigma Chemical Company, St. Louis, Mo.) were added to agar and broth media when appropriate: ampicillin, 100 µg/ml; tetracycline, 10 µg/ml; and kanamycin, 50 µg/ml. Chemicals in buffers and media were obtained from Sigma.

General genetic methods

Procedures for the isolation of genomic DNA and transformation were carried out as described by Sambrook et al. Plasmid DNA was isolated using the QIAprep Spin Minikit and protocol (Qiagen Inc., Chatsworth, Calif.). Restriction and DNA-modifying enzymes were used as recommended by the manufacturer (New England Biolabs, Beverly, Mass.). DNA fragments were compared to a 1-kb DNA ladder (Promega Corp., Madison, Wis.) and then purified from 1% agarose gels (Gibco BRL, Grand Island, N.Y.) using the Geneclean II kit (Bio 101, Inc., Vista, Calif.). Primary DNA cloning and manipulation were conducted with *E. coli* DH5. PCR amplification of DNA was performed using an Amplitron II Thermal Cycler (Barnstead/Thermolyne, Dubuque, Iowa), and conditions were denaturation at 94°C for 1 min, annealing at 60°C for 1 min, and extension at 72°C for 2 min for 30 cycles. All reagents and *Taq* DNA polymerase were used as recommended by the manufacturer (Perkin-Elmer Co., Foster City, Calif.).

Construction of dps::nptI suicide vector

The *dps* gene in pSC9911 was inactivated in vitro by insertion of *nptI* encoding aminoglycoside 3'-phosphotransferase and confers resistance to kanamycin. The 1.2-kb DNA fragment carrying *nptI* was isolated from pUC4K and digested with *Pst*I, and the resulting cohesive termini were converted to blunt ends with Klenow fragment. The *nptI* fragment was inserted into a unique *Nco*I site present within the open reading frame (ORF) of *dps*. The resulting construct (pSC9921) containing the 2.0-kb *dps::nptI* fragment was digested with *Eco*RI and *Pst*I to liberate the *dps::nptI* cartridge. The DNA fragment was blunt ended with Klenow fragment and ligated with *Sma*I-digested pCVD442, forming pSC9922. pCVD442 is a suicide vector containing the R6K origin of replication that requires the protein in *trans* encoded by *pir*. *E. coli* SY327 *pir* was transformed with pSC9922. The plasmid pSC9932, which was identical to pSC9922 except that the *dps::nptI* cartridge was in the opposite orientation, was constructed by following the procedures for pSC9922.

Generation of the dps::nptI mutant

The suicide vectors, pSC9922 and pSC9932, were used to generate the *dps::nptI* mutants (FRIK 47991 and FRIK 47992) in *E. coli* O157:H7 strain ATCC 43895 by homologous recombination. Both plasmids contain the RP4 origin of transfer (*oriT*) and can be conjugally mobilized from donor cells containing the *tra* gene. Therefore, *E. coli* SM10 *pir tra* was transformed separately with pSC9922 and pSC9932 and used as a conjugal donor to *E. coli* O157:H7 strain ATCC 43895. Conjugation was conducted using methods previously described with some modification. The recipient strain ATCC 43895 and donor strain SM10 *pir tra* (containing pSC9922 or pSC9932) were grown overnight on LB agar, removed with a sterile cotton swab, spotted on LB agar, and mixed thoroughly. The donor-recipient mixture was incubated at 37°C for 8 h and then resuspended in 1 ml of saline (0.85% NaCl). Portions of the cell mixture (100 μl) were spread on each of 10 plates of MSA supplemented with kanamycin and sucrose (6%) and incubated overnight at 37°C.

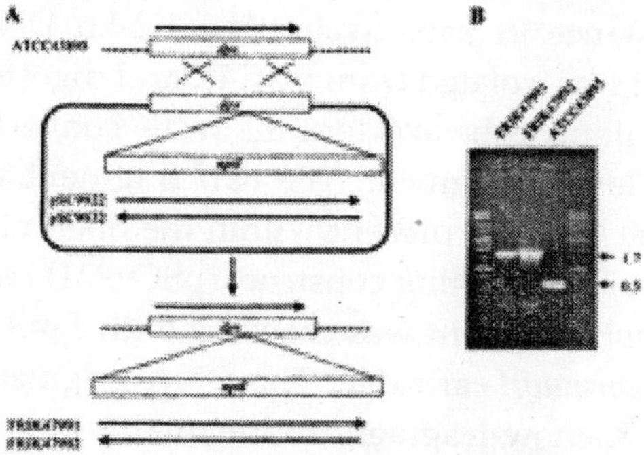

FIG. / Diagram of allelic exchange and confirmation of the dps::nptI mutants. (A) Homologous recombination between chromosomal dps gene from strain ATCC 43895 and pSC9922 or pSC9932. Dashed lines, chromosomal DNA; solid line, plasmid DNA; dps boxes, the target dps gene; nptI boxes, the nptI gene; ×, crossover. (B) PCR analysis of E. coli O157:H7 dps mutant generated by allelic exchange (FRIK 47991 and FRIK 47992). Strain ATCC 43895 is the parent strain for FRIK 47991 and FRIK 47992. Molecular size markers (1-kb ladder [Promega Corp.]) appear in the end lanes of the gel. The oligonucleotides for PCR amplification are presented in Materials and Methods.

The desired transconjugants were selected by their ability to grow on MSA supplemented with kanamycin and sucrose. Strains that were kanamycin resistant and ampicillin sensitive, to confirm the absence of the suicide vector (pSC9922 or pSC9932), were selected for further study. Selected strains were then confirmed for the O157 antigen by slide agglutination (Oxoid, Basingstoke, England) and for the presence of *nptI* in the *dps* gene by PCR amplification. The primers, *dps*kk-1 (5′GCCGGAATTCATGAAATTATGAGTACCGC3′) and *dps*kk-2 (5′GAAAACTGCAGAATTTATCCAGGTCGCGAGACGACGC3′) were designed to hybridize and amplify the *dps* ORF. The PCR end products were analyzed by agarose gel electrophoresis to confirm the presence of *nptI* in *dps* (1.7-kb fragment).

Complementation of the dps::nptI mutant

The *dps* gene (790 bp) and promoter from *E. coli* O157:H7 strain ATCC 43895 were amplified by PCR and cloned into the *Eco*RI site of pBR322 to generate pSC9911. The *tet* gene of pSC9911 was then removed using *Hin*dIII and *Ava*I, treated with Klenow fragment, and blunt end ligated to produce pSC9915 (Table 1). pSC9916 was used as a control and was generated from pBR322 by removing the *tet* gene. Each plasmid (pSC9915 and pSC9916) was transformed into FRIK 47991 (*dps::nptI*) using standard methods and used in subsequent acid and hydrogen peroxide challenge studies.

Acid and hydrogen peroxide challenges

Acid tolerance was assessed in synthetic gastric fluid adjusted to pH 1.8 with HCl (12 N) and filter sterilized as described previously. Synthetic gastric fluid was prepared essentially as described by Beumer et al. except that bovine bile was used in place of porcine bile. Cultures in exponential and stationary phases of growth were used to inoculate flasks containing 100 ml of synthetic gastric fluid to achieve a final concentration of ca. 10^5 CFU/ml.

Following inoculation, the flasks were incubated at 37°C with shaking (150 rpm) and samples were removed at appropriate intervals, plated in duplicate on tryptic soy agar using a Model D Spiral Systems plater (Cinncinati, Ohio), and incubated at 37°C. The percent survivors was calculated using the CFU per milliliter as determined immediately after inoculation as 100%. The limit of detection for this method was 10 CFU/ml (12); hence, a maximum decrease of 4 $\log_{10}$ CFU/ml could be detected. Additionally, colonies were randomly tested for each respective phenotype by two methods. The first was by growth in the presence of the respective antibiotic(s), and the second was by PCR amplification of the ***dps::nptI*** cartridge and *dps* complement when applicable.

The assay for survival of log- and stationary-phase cells in the presence of 15 mM hydrogen peroxide was conducted as previously described.

Hydrogen peroxide was added to cell suspensions (ca. 10^5 CFU/ml) in 100 ml of phosphate-buffered saline (0.01 M, pH 7.2) and incubated at room temperature (22°C) with shaking (100 rpm). Samples were removed periodically to determine the number of CFU per milliliter as described for acid challenges. All D values were calculated using the formula (D_{value} = 1/ slope), where slope represents the linear regression of the data including first and last points.

Cloning and sequencing of dps

A DNA fragment containing the *dps* structural gene and upstream regulatory region was amplified from genomic DNA of *E. coli* O157:H7 strain ATCC 43895 by PCR using a pair of oligonucleotide primers carrying *Eco*RI or *Pst*I sites on the 5' ends. The primers (*dps*-1, 5'CGGAATTCCATAACCA-TGCAGAATTTCT3', sense primer, and *dps*-2, 5'CGGCTGA-GCAGCGATGGATTTATTCGAT3', antisense primer) were designed using the *dps* sequence of *E. coli* K-12 (GenBank accession no. X69337) and synthesized (Gibco BRL, Gaithersburg, Md.).

The resulting PCR fragment was digested with *Eco*RI and *Pst*I and ligated into pBR322, previously digested with the same enzymes, to produce pSC9911. The nucleotide sequence of the 790-bp DNA fragment in pSC9911 was determined (University of WisconsinMadison Biotechnology Center). Sequence and amino acid comparisons were conducted using BLAST (National Center for Biotechnology Information).

Statistical analyses

The data reported are the average values from three trials and were analyzed using the *t* test with SigmaStat (Jandel Scientific, San Rafael, Calif.) software.

Nucleotide sequence accession number

The nucleotide sequence of *dps* from *E. coli* O157:H7 strain ATCC 43895 was deposited in the GenBank database under accession no. AF140030.

RESULTS

Generation and confirmation of the dps::nptI mutant in E. coli O157:H7

Transconjugants resulting from the conjugation of pSC9922 from SM10 *pir* to O157:H7 strain ATCC 43895 were kanamycin resistant, ampicillin sensitive, and sucrose positive. Strains selected for further study were also positive for the O157 antigen. While allelic exchange between the insert *dps::nptI* and chromosomal *dps* can occur by a double crossover, *sacB* in pSC9922 and pSC9932 encodes levansucrase and selects against the maintenance or integration of these plasmids into the chromosome. Confirmation of a double crossover in which wild-type *dps* was replaced with the *dps::nptI* allele was confirmed by PCR.

PCR analysis of genomic DNA from parental strain ATCC 43895 with primers *dps*kk-1 and *dps*kk-2 produced a 481-bp fragment, whereas genomic DNA from the *dps::nptI* mutant(s) resulted in an amplified DNA fragment approximately 1.7 kb in length. The 1.7-kb fragment is in agreement with the projected size of the DNA fragment containing wild-type *dps* (481 bp) and the *nptI* gene (1.2 kb). FRIK 47991.

Acid tolerance

The survival of log-phase cells (5 h, $A_{600} = 0.8$) of the parent strain (ATCC 43895) was significantly greater ($P < 0.05$) than that of the *dps::nptI* mutant (FRIK 47991) when challenged in synthetic gastric fluid (pH 1.8). The $D_{\mathrm{pH}\,1.8}$ values for the parent and mutant strains were 157 and 36 min, respectively. The survival of *dps::nptI* mutants, FRIK 47991 and FRIK

47992, containing *nptI* in the opposite orientation, did not differ significantly in acid challenges (data not shown). FRIK 47991 was chosen for further study. Complementation of *dps::nptI* in FRIK 47991 with a functional *dps* gene (pSC9915) restored acid tolerance to a level equivalent to that of the parent strain. FRIK 47991 containing the control plasmid (pSC9916) displayed survival that was significantly impaired ($P < 0.01$) compared to that of the parent strain and FRIK 47991 harboring pSC9915.

Similar to results with log-phase cells, the survival of early-stationary-phase cells (12 h, $A_{600} = 1.1$) of the parent strain and FRIK 47991 containing pSC9915 was significantly greater ($P < 0.05$) than that of FRIK 47991 and FRIK 47991 containing the control plasmid pSC9916. The $D_{pH\ 1.8}$ values were 93 min for the parent strain and 71 min for the complemented strain (FRIK 47991 with pSC9915) compared to 49 and 45 min for FRIK 47991 with and without the control plasmid (pSC9916), respectively.

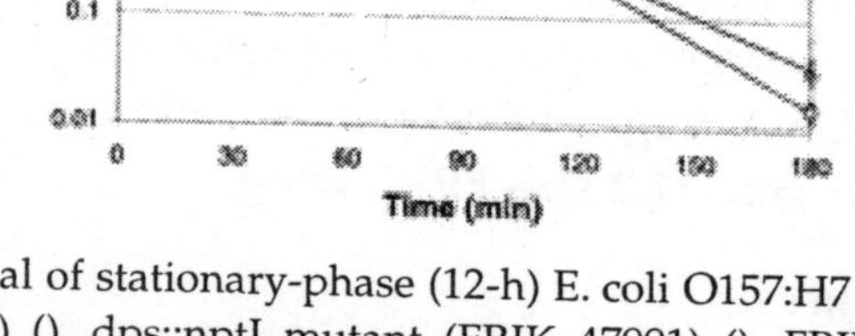

FIG. 3. Survival of stationary-phase (12-h) E. coli O157:H7 parent strain (ATCC 43895) (), dps::nptI mutant (FRIK 47991) (), FRIK 47991 with pSC9915 (), and FRIK 47991 with pSC9916 () in synthetic gastric fluid (pH 1.8). All points represent the means from three independent trials. Error bars represent the standard errors.

The trends in survival of the late-stationary-phase cells were similar to those for log-phase and early-stationary-phase cells with the exception of the *dps::nptI* mutant (FRIK 47991).

The survival of the parent strain, FRIK 47991, and FRIK 47991 with either pSC9915 or pSC9916 was equivalent to or greater than the survival of log- and early-stationary-phase cells. FRIK 47991 containing pSC9915 exhibited the lowest rate of decline and had the greatest number of survivors after 4 h of acid challenge.

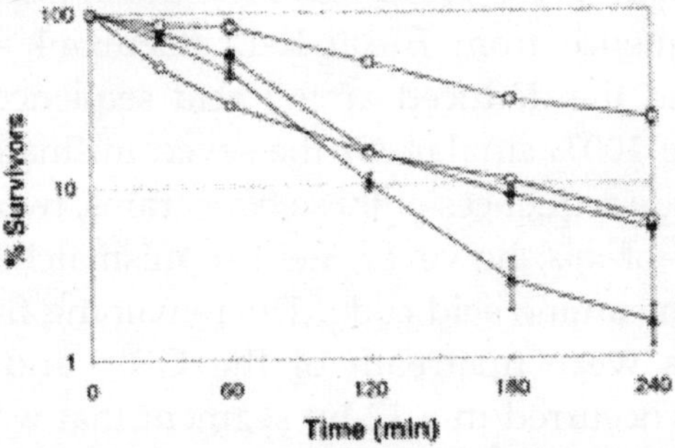

FIG. 4. Survival of late-stationary-phase (24-h) E. coli O157:H7 parent strain (ATCC 43895) (), dps::nptI mutant (FRIK 47991) (), FRIK 47991 with pSC9915 (), and FRIK 47991 with pSC9916 () in synthetic gastric fluid (pH 1.8). All points represent the means from three independent trials. Error bars represent the standard errors.

Survival during hydrogen peroxide challenge

The survival of the parent strain was significantly greater ($P < 0.05$) than that of FRIK 47991 when challenged in 15 mM H_2O_2 regardless of the growth phase. The maximum survival reported as D values for all strains was with early-stationary-phase cells (12 h). As reported for acid challenges, complementation of *dps::nptI* with a functional *dps* gene (pSC9915) restored hydrogen peroxide tolerance to levels equal to or greater than that of the parent strain. The presence of pSC9916 (control plasmid) in FRIK 47991 improved survival slightly, but D values were still significantly less ($P < 0.1$) than those for the parent strain.

Cloning and nucleotide sequence of dps

A 790-bp DNA fragment containing the *dps* structural gene and upstream regulatory region was amplified by PCR from genomic DNA of *E. coli* O157:H7 strain ATCC 43895. The nucleotide sequence of the amplified fragment was determined (accession no. AF140030). An ORF starting at nucleotide 276 and ending at nucleotide 779 was identified. The *dps* gene codes for a 167-amino-acid protein with an estimated mass of 18.7 kDa and a pI of 5.72. The nucleotide sequence of the *dps* from O157:H7 strain ATCC 43895 was 99% similar (783 of 790 bp) to the *dps* sequence from *E. coli* K-12 (GenBank accession no. X69337), and the deduced amino acid sequences from these strains were 100% similar. Of the seven mismatches between the nucleotide sequences of these two strains, two were located in the ORF of *dps*; however, neither mismatch resulted in a change in the amino acid code. The remaining five nucleotide mismatches were upstream of the ORF, and four of the mismatches occurred in a 12-bp segment that was flanked by inverted repeats (data not shown).

DISCUSSION

Bacteria have evolved with elaborate protection systems to allow survival and/or growth during exposure to acidic environments. The three main defense strategies which protect the cell from acid are changes in membrane composition, homeostasis systems for internal pH and pathways for repair-protection of essential cellular components. Studies with *E. coli*, *Salmonella*, and *H. pylori* have suggested that DNA repair pathways are important for survival in low-pH conditions such as the gastric barrier. Almirón et al. identified a DNA-binding protein (Dps) that is regulated by ^{38}, ^{70}, and OxyR and protects DNA from damage. Dps forms dodecamers, analogous to ferritins, that interact with DNA to form a stable complex that protects DNA from the hydroxyl radicals formed during oxidative stress. Low pH also damages DNA as a result of selective depurination reactions and can cause lesions in double-stranded DNA. This study investigated the role of Dps in the acid stress and oxidative stress tolerance of *E. coli* O157:H7, which epidemiological and laboratory studies demonstrate is particularly tolerant of low-pH conditions.

A *dps::nptI* mutant (FRIK 47991) was generated and used to determine if *dps* contributes to acid tolerance in *E. coli* O157:H7. During our evaluation of plasmid constructs to complement the *dps::nptI* mutant, acid tolerance of the *dps* mutant FRIK 47991 was further reduced when transformed with pSC9911 (data not shown). In follow-up experiments, the introduction of pBR322 into the parent strain *E. coli* O157:H7 also decreased acid tolerance (data not shown) and indicated that *tetA*, which encodes a tetracycline pump, decreased acid tolerance (C.-M. Cheng, J. L. Bose, S. H. Choi, and C. W. Kaspar, unpublished observation). Previous studies have documented that the tetracycline resistance gene (*tetA*) causes pleiotropic effects in addition to directing the efflux of tetracycline from the bacterium. Therefore, the *tetA* gene was removed from pSC9915 (Ap^r Tc^s; used to complement *dps::nptI* in FRIK 47991) and pSC9916 (Ap^r Tc^s; control plasmid).

Characterization of the survival properties of the parent strain (ATCC 43895) and the *dps::nptI* mutant (FRIK 47991) demonstrated significant differences during acid and hydrogen peroxide challenges, particularly when log-phase (5-h) cultures were examined. Although *dps* has been primarily characterized in stationary-phase protection, these results were not unexpected because *dps* is regulated by ^{38}, ^{70}, and OxyR. This is also consistent with the growth phase variations in nucleoid composition in log-phase cells observed by Azam et al. Complementation of the *dps::nptI* mutant with pSC9915 and the restoration of acid stress and oxidative stress tolerance to levels equivalent to or exceeding that of the parent strain demonstrates that Dps or a Dps-regulated protein is responsible for the observed differences in survival.

Early-stationary-phase (12-h) cultures of the parent strain were significantly more tolerant of acid than was the *dps::nptI* mutant. In both log-phase and 12-h cultures, the parent strain decreased by 1 to 2 $\log_{10}$ CFU/ml after 3 h of acid challenge while FRIK 47991 decreased by ca. 4 $\log_{10}$ CFU/ml. In contrast to the results in acid challenges, 12-h cultures of FRIK 47991 were significantly more tolerant of hydrogen peroxide challenge (ca. 2.5-$\log_{10}$-CFU/ml reduction after 60 min) than were log-phase cultures (4-$\log_{10}$-CFU/ml reduction after 60 min), although the survival of the parent strain was still significantly greater ($P < 0.01$) than that of FRIK 47991. However, the final numbers (CFU per milliliter) of the parent and *dps::nptI* mutant after 2 h of hydrogen peroxide challenge were essentially the same (data not shown).

Stationary-phase cells of the parent strain also exhibited increased tolerance for hydrogen peroxide challenge in comparison with log-phase cultures. The increased tolerance of stationary-phase cells for hydrogen peroxide challenge is likely due to the production of catalase. *E. coli* O157:H7 possess three separate catalase genes (*katG, katE,* and *katP*). *katG* is regulated by OxyR and predominantly produced during log phase, while *katE* is regulated by *rpoS* and primarily expressed in stationary phase. The regulation of the plasmid-encoded (pO157) catalse (*katP*) has not been elucidated. Thus, stationary-phase production of catalase (*katE*) likely provides protection against hydrogen peroxide and compensates for the absence of Dps.

The survival of late-stationary-phase cells (24 h) of the parent strain in acid challenges was similar to that observed with log-phase (5-h) and early-stationary-phase (12-h) cells. There was a significant improvement in the acid tolerance of the *dps::nptI* mutant in late-stationary-phase cells as viable numbers decreased less than did log- and early-stationary-phase cultures. In fact, the numbers of FRIK 47991 survivors were equivalent to those of the parent strain after 2 h of acid challenge, although there were greater numbers of the parent strain recovered after 30 and 60 min of acid challenge (Fig. 4).

The production of other proteins with protective roles in stationary phase, including the DNA-binding proteins encoded by *cbpA* and *rob*, most likely provides protection from acid in the absence of Dps. The results from hydrogen peroxide challenges with late-stationary-phase cells of the parent strain were similar to those obtained with log-phase cells. The *D* values from hydrogen peroxide challenges with FRIK 47991 in log phase, early stationary phase, and late stationary phase were 15, 35, and 19 min, respectively. Evidently, the proteins that enhanced acid tolerance in late-stationary-phase cells of FRIK 47991 are less effective in protecting the bacterium from oxidative stress.

As noted above, the complementation of the *dps::nptI* mutant (FRIK 47991) with pSC9915 increased survival in acid and hydrogen peroxide challenges to a level equivalent to or exceeding that of the parent strain (ATCC 43895). The increased tolerance can be attributed to the multiple copies of *dps* provided by pSC9915 compared to the single-chromosome-encoded allele found in the parent strain. Transformation of FRIK 47991 with the control plasmid pSC9916 did not restore acid and hydrogen peroxide tolerance; however, in challenges with log-phase cells, FRIK 47991 containing pSC9916 survived better than did FRIK 47991 without the control plasmid. It is possible that the presence of multiple copies of pSC9916 provided some protection against acid and hydrogen peroxide in log-phase cells. This may be explained by the presence of additional nonessential DNA targets that would decrease the rates of depurination from low pH and oxidation of sugar and base moieties from oxygen radicals that target chromosomal DNA.

The sequence homology of *dps* from *E. coli* O157:H7 strain ATCC 43895 and the fact that the deduced amino acid sequences were identical to *E. coli* K-12 demonstrate that the acid tolerance noted for some serotype O157:H7 strains is not due to differences in the ORF of this gene. Studies are in progress to determine if the inverted rêpeat and nucleotide differences in the upstream region of *dps* in *E. coli* O157:H7 influence regulation.

Results from this study demonstrate that *dps* makes a significant contribution to the acid tolerance of *E. coli* O157:H7. In addition to acid tolerance, Dps is important in oxidative stress protection as reported previously for non-serotype O157:H7 *E. coli*. It is likely that Dps protects DNA from the deleterious effects of low pH in a manner analogous to oxidative stress protection; however, it is possible that Dps influences expression of other genes that protect or repair DNA or provide acid tolerance by another mechanism. Regardless, Dps is a key component of the general stress protection system that is important in the survival of the bacterium.

6

Inactivation of Gram - Negative Bacteria by Lysozyme

We have studied the inactivation of six gram-negative bacteria (*Escherichia coli, Pseudomonas fluorescens, Salmonella enterica* serovar Typhimurium, *Salmonella enteritidis, Shigella sonnei,* and *Shigella flexneri*) by high hydrostatic pressure treatment in the presence of hen egg-white lysozyme, partially or completely denatured lysozyme, or a synthetic cationic peptide derived from either hen egg white or coliphage T4 lysozyme. None of these compounds had a bactericidal or bacteriostatic effect on any of the tested bacteria at atmospheric pressure.

Under high pressure, all bacteria except both *Salmonella* species showed higher inactivation in the presence of 100 μg of lysozyme/ml than without this additive, indicating that pressure sensitized the bacteria to lysozyme. This extra inactivation by lysozyme was accompanied by the formation of spheroplasts.

Complete knockout of the muramidase enzymatic activity of lysozyme by heat treatment fully eliminated its bactericidal effect under pressure, but partially denatured lysozyme was still active against some bacteria. Contrary to some recent reports, these results indicate that enzymatic activity is indispensable for the antimicrobial activity of lysozyme. However, partial heat denaturation extended the activity spectrum of lysozyme under pressure to serovar Typhimurium, suggesting enhanced uptake of partially denatured lysozyme through the serovar Typhimurium outer membrane. All test bacteria were sensitized by high pressure to a peptide corresponding to amino acid residues 96 to 116 of hen egg white, and all except *E. coli* and *P. fluorescens* were sensitized by high pressure to a peptide corresponding to amino acid residues 143 to 155 of T4 lysozyme. Since they are not enzymatically active, these peptides probably have a different mechanism of action than all lysozyme polypeptides.

INTRODUCTION

High hydrostatic pressure treatment is a promising technique for cold pasteurization of foods that allows better retention of product flavor, texture, color, and nutrient content than a comparable conventional heat pasteurization. The main obstacles that prevent a commercial breakthrough of pressure-preserved foods are the high investment cost, due to the high pressures required for efficient microbe and enzyme destruction, and the paucity of knowledge on the sensitivity of various pathogenic and spoilage microorganisms to hydrostatic pressure and on the factors affecting this sensitivity.

The application of hurdle technology has been proposed as an approach to increase the microbicidal effect of the process at lower pressures. Hurdle technology relies on the synergistic combination of moderate doses of two or more microbe-inactivating and/or growth-retarding factors. An interesting example of synergistic inactivation exists between high pressure and a number of antimicrobial peptides, including nisin, lysozyme, and pediocin.

This synergistic inactivation was observed not only in intrinsically sensitive gram-positive bacteria but also in gram-negative bacteria, which are normally insensitive to these peptides because their cellular targets are shielded by their outer membrane.

In the present work we have focused on lysozyme because it has some interesting features for application as a food preservative. First, lysozymes are naturally present in foods such as egg white (ca. 3.2 mg/ml), cow milk (ca. 0.13 µg/ml), and human colostrum (ca. 65 µg/ml) but also in several plants, such as cauliflower (ca. 27.6 µg/ml) and cabbage (ca. 2.3 µg/ml). The use of these naturally occurring lysozymes at a concentration of 10 to 100 µg/ml, as proposed in this work, should therefore not present a toxicological concern. Second, the bacteriostatic and bactericidal properties of lysozyme have been the subject of many studies, and over the last 10 years, several authors have proposed a novel antibacterial mechanism of action for lysozyme that is independent of its 1,4--*N*-acetylmuramidase activity. These reports are based on unique bactericidal properties observed with partially or completely denatured lysozymes having reduced or no enzymatic activity against both gram-positive and gram-negative bacteria. For example, Ibrahim et al. demonstrated inactivation by heat-denatured lysozyme of an *Escherichia coli* K-12 strain, which was relatively insensitive to native lysozyme. Düring et al. showed that native and heat-treated enzymatically inactive lysozyme from coliphage T4 caused similar inactivation levels on a particular *E. coli* strain. The antibacterial properties of denatured lysozyme have been proposed to result from the cationic nature of the peptide in combination with conformational changes leading to increased hydrophobicity. These characteristics are believed to contribute to the antimicrobial properties of several peptides. Specific peptides with cationic properties and without enzymatic activity derived from hen egg white and T4 lysozyme were also found to have antimicrobial activity.

The present work extends our previous observation that high pressure sensitizes *E. coli* for lysozyme. We have studied the effect of denatured lysozymes and lysozyme-derived peptides, in addition to native lysozyme, in combination with high pressure on a panel of six different gram-negative bacteria including several foodborne pathogens. The results of this work should provide better insight into the mode of antibacterial action of lysozyme and contribute to the development of more efficient technology for cold high-pressure pasteurization.

MATERIALS AND METHODS

Bacterial cultures and growth conditions. The bacteria used in this work are *E. coli* K-12 strain MG1655, *Pseudomonas fluorescens* LMMBM07, *Salmonella enterica* serovar Typhimurium LMMBM01 (both from our laboratory collection), *Salmonella enteritidis* ATCC13076 (from the American Type Culture Collection), *Shigella flexneri* LMG10472 and *Shigella sonnei* LMG10473 (both from the Belgian Coordinated Culture Collection of Microorganisms, Ghent, Belgium). All experiments were carried out with cultures in stationary phase, obtained by growth in nutrient bouillon (Oxoid, Basingstoke, United Kingdom) for 21 h with shaking (200 rpm) at 37°C, except for *P. fluorescens*, which was incubated at 30°C.

Growth inhibition. Growth inhibition by lysozymes and lysozyme-derived peptides was determined by recording growth curves in triplicate with a Bioscreen C microbiology reader (Labsystems Oy, Helsinki, Finland). Stationary phase cultures were diluted to between 5×10^3 and 5×10^4 cells/ml in fresh medium, and 380 μl was transferred to the honeycomb plate wells of the Bioscreen C reader. The volume was then adjusted to 400 μl, either with buffer for the controls or with the appropriate solution of antimicrobial. Every 15 min the cultures were shaken at medium intensity for 1 min, and the turbidity was measured with a wide band filter. The growth curves were followed for 30 h, in which time all bacteria reached the stationary phase.

Denaturation of lysozyme and measurement of lysozyme enzymatic activity. Hen egg-white lysozyme (66,000 U/mg;

Fluka, Buchs, Switzerland) was stored frozen (20°C) as a stock solution of 1 mg/ml in potassium phosphate buffer (10 mM, pH 7.0). For heat denaturation, 100 µl of the 1-mg/ml solution was transferred to a sterile glass capillary and treated for 20 or 60 min at 80 or 100°C in a water bath. For denaturation with -mercapto-ethanol, the 1-mg/ml lysozyme solution was incubated with 5% -mercapto-ethanol during 1 h at 60°C. After the treatments, the lysozyme samples were put directly on ice and stored at 20°C.

The enzymatic activity of lysozyme and its derivatives was measured with lyophilized *Micrococcus lysodeikticus* (ATCC 4698) cells (Fluka) resuspended at 0.5 mg/ml in 10 mM potassium phosphate buffer (pH 7.0) as a substrate using a method adapted from Weisner. Thirty-microliter aliquots of different dilutions of the sample were added to 300 µl of *M. lysodeikticus* cell suspension, and the lysis of cells was measured automatically as the decrease in turbidity (optical density at 600 nm [OD_{600}]) with the Bioscreen C microbiology reader during 40 min at 20°C. The dilution resulting in a rate of turbidity decrease between 0.001 and 0.005 OD U/min was used to calculate the enzymatic activity. Enzymatic activity was expressed in units per milligram of protein or as a percentage relative to untreated lysozyme.

Synthetic peptides. Two synthetic peptides (95% purity) were purchased from Eurogentec (Herstal, Belgium). Peptide HEL96-116 (H_2N-KKI VSD GNG MNA WVA WRK RCK-COOH) is a 21-mer peptide corresponding to amino acids 96 to 116 of hen egg-white lysozyme (HEL). Peptide T4L143-155 (H_2N-PNR AKR VIT TFR T-COOH) is a 13-mer peptide and corresponds to amino acids 143 to 155 of bacteriophage T4 lysozyme (T4L).

Pressure treatment. Cells in stationary phase were harvested by centrifugation (3,800 × *g*, 5 min) and resuspended in the same volume of potassium phosphate buffer (10 mM, pH 7.0), yielding a final cell population of 5×10^8 to 5×10^9 CFU/ml. Although the pH of phosphate buffer is more pressure dependent than the pH of some other buffers, phosphate buffer was chosen for this study because it is widely used in inactivation studies and is not harmful to microorganisms. After the addition of lysozyme or one of its derivatives where appropriate, cell suspensions (300 µl) were sealed without air bubbles in sterile polyethylene bags and subjected at 20°C to pressures in the range of 155 to 300 MPa. Pressure treatment was done in a system with eight parallel thermostatically controlled 8-ml vessels which could be simultaneously pressurized and individually decompressed at different times (Resato, Roden, The Netherlands). The compression rate was approximately 100 MPa/min; decompression was in less than 3 s. The high-pressure transmission fluid used was Resato high-pressure fluid TP1, a mixture of glycols (Van Meeuwen, Wesp, The Netherlands). The pressurization times reported do not include the come-up and come-down time. It should also be noted that the temperature in the vessels could not be kept constant, due to adiabatic compression and decompression. Temperature measurements with thermocouples inside the pressure vessels, which had been previously conducted under identical circumstances, suggested a temperature increase to 29°C upon rapid pressurization to 300 MPa.

To measure bactericidal activities at atmospheric pressure, a part of the suspension with the additives was kept at room temperature without pressurization and was plated after the same exposure time as the pressurized samples.

Enumeration of viable cells. Appropriate dilutions in sterile potassium phosphate buffer (10 mM, pH 7.0) were surface plated with a spiral plater (Spiral System Inc., Cincinnati, Ohio) on tryptic soy agar for *E. coli* and *P. fluorescens* or on plate count agar (both media from Oxoid) for the other bacteria. The plated volume was 50 µl, and hence the detection limit was 20 CFU/ml. Colonies were allowed to develop for 24 to 48 h at the appropriate incubation temperature. Inactivation was expressed as a logarithmic viability reduction, log (N_0/N), with N and N_0 the colony counts after a treatment and in the untreated sample, respectively. For all treatments, averages ± standard deviations for at least three independent cultures of each strain are shown. Significant differences were calculated with the paired Student's *t* test.

RESULTS

Sensitization for native lysozyme. The test panel of six gram-negative bacteria including four pathogens (two *Salmonella* and two *Shigella* strains) was screened for sensitivity to native lysozyme under conditions of ambient and of elevated pressure. Two concentrations, 10 and 100 µg of lysozyme/ml, were used to investigate dose dependency. Growth curves in the presence of lysozyme revealed no inactivation or growth retardation for any of the bacteria, even at 100 µg of lysozyme/ml (data not shown). Higher concentrations of lysozyme were not tested, since these tended to cause aggregation of the bacteria, making the plate counts unreliable.

Sensitization for lysozyme by high pressure was tested by adding lysozyme to the bacterial suspensions before pressure treatment. For each strain a pressure was chosen that, in the absence of lysozyme, caused an inactivation of at least 1 log unit. We speculated that in this way the pressure treatment would be severe enough to sensitize the cells to lysozyme. Because of their different pressure sensitivities, a uniform pressure treatment could not be used for all the bacteria. At 10 µg/ml, only *E. coli* and *P. fluorescens* were sensitized to lysozyme by high pressure, but at 100 µg/ml, *S. flexneri* and *S. sonnei* also became sensitive, and the extra logarithmic viability reduction caused by lysozyme for *E. coli* and *P. fluorescens* increased from 0.4 to 1.0 and from 1.5 to 2.5, respectively. Neither *Salmonella* strain was inactivated by lysozyme under pressure, and the presence of 100 µg of lysozyme/ml even had a protective effect against pressure inactivation for *Salmonella* serovar Typhimurium. In three treatments (*S. sonnei* with lysozyme at 10 and 100 μg/ml, and serovar Typhimurium at 100 μg/ml), standard deviations were remarkably higher than in any other treatment. These experiments were repeated in triplicate but the standard deviations remained high.

Sensitivity for lysozyme under pressure was transient, since exposure of pressure-treated cells to lysozyme (1 h at room temperature) after pressure treatment did not cause further inactivation, not even for *P. fluorescens*, which was most sensitive for lysozyme under pressure.

Sensitization for partially heat- or -mercapto-ethanol-denatured lysozyme. It has been suggested by some authors that partial or even complete heat denaturation would extend the working spectrum of lysozyme to some gram-negative bacteria that are not normally sensitive to lysozyme. We wanted to confirm these remarkable observations on our own panel of test bacteria and to compare the bactericidal efficiency of native and denatured lysozymes under high pressure. First, we tested whether lysozyme is denatured by the high pressure treatment itself. This seems not to be the case, since even after a harsh treatment at 600 MPa and 60°C during 15 min, 100% of the enzymatic activity was retained (Table 2). Lysozyme was denatured with heat in a way similar to that described by Ibrahim et al. and, in addition, with -mercapto-ethanol.

The different treatments of lysozyme, the designations used for each denatured product, and the corresponding remaining enzymatic activities are shown in Table 2. Two denatured forms (H100/20-lys and M-lys) were almost completely enzymatically inactive with, respectively, 0.6 and 0.5% of residual activity whereas one (H80/20-lys) still had 11.5% of lytic activity. In the high pressure experiments, H80/20-lys was applied at 10 μg/ml and the more denatured forms at 10 and 100 μg/ml.

In control experiments at atmospheric pressure, no bacteriostatic or bactericidal activity could be detected for any of the denatured forms on any of the test bacteria (data not shown). Since the solution of M-lys still contained -mercapto-ethanol, another series of control experiments was performed in which all test bacteria were pressure treated in the presence of 5% -mercapto-ethanol, but no enhanced lethality was observed due to the presence of this compound (data not shown). Table 3 shows the reduction factors of all the test bacteria when treated with the denatured lysozymes under high pressure. In general, denatured lysozymes were active under pressure against fewer bacteria than intact lysozyme at the same concentration.

Both strongly denatured lysozymes (H100/20-lys and M-lys) were completely inactive under pressure at 10 µg/ml, although they remained active against at least one of the bacteria at 100 µg/ml. An interesting observation concerns serovar Typhimurium, which was sensitive under pressure for some of the denatured lysozymes (H80/20-lys at 10 µg/ml and H100/20-lys at 100 µg/ml), while it was not sensitive for native lysozyme even at 100 µg/ml. This remarkable behavior of serovar Typhimurium was confirmed in an experiment with six replicate samples. Serovar Typhimurium was not sensitive, however, to the -mercapto-ethanol-denatured lysozyme (M-lys), although the latter had almost the same level of enzymatic activity as H100/20-lys. M-lys at 100 µg/ml was active only against *P. fluorescens*. This organism was also sensitive to H80/20-lys (10 µg/ml) and H100/20-lys (100 µg/ml) under pressure and was therefore the most sensitive of the tested bacteria.

Contribution of enzymatic activity to the bactericidal properties of lysozymes under pressure. In the experiment described above, the denatured lysozymes always had some residual enzymatic activity. To clarify whether enzymatic activity is necessary at all for the bactericidal effects observed with denatured lysozymes under pressure, we subjected lysozyme to heat treatment at 100°C during 60 min, obtaining a sample with undetectable enzymatic activity (H100/60-lys). The inactivation of *P. fluorescens*, as the most sensitive of all tested bacteria in the previous experiments, by H100/60-lys and H80/20-lys under pressure was subsequently compared and it was found that the completely denatured lysozyme had lost its bactericidal activity (data not shown). A further confirmation for the role of peptidoglycanolytic activity was found by microscopic observation of cell morphology after pressure treatment. The bactericidal effect caused by intact or partially denatured lysozyme under high pressure was always accompanied by a change in the morphology of the cells from rod to sphere. This change did not occur in the absence of lysozyme or with the completely denatured lysozyme and is therefore most likely due to the formation of spheroplasts as a result of the residual lytic activity of lysozyme.

Sensitization for lysozyme-derived peptides. In a final set of experiments, two synthetic peptides derived, respectively, from hen egg-white lysozyme (HEL96-116) and *E. coli* bacteriophage T4 lysozyme (T4L143-155) were investigated for antibacterial activity. Peptide HEL96-116 is similar to the bactericidal peptide of 15 amino acids that was isolated by Pellegrini et al. by digesting lysozyme with the protease clostripain but has two additional NH_2-terminal and four additional COOH-terminal amino acids from the original lysozyme sequence. This increases the cationic character of the peptide, which is known to contribute to the antibacterial activity of several antibiotic peptides. Peptide T4L143-155 was chosen and synthesized by Düring et al. for its amphiphatic character and helicoidal structure and was also found to have bactericidal activity. The calculated isoelectric points of HEL96-116 and T4L143-155 are 10.29 and 12.40, respectively, and both peptides were completely enzymatically inactive.

These two peptides were applied to our test panel of bacteria at 100 μg/ml. At atmospheric pressure neither growth inhibition nor inactivation was observed (data not shown). Under pressure, all bacteria were sensitized for the HEL96-116 peptide, even serovar Typhimurium and *S. enteritidis*, which were not sensitized for native lysozyme under pressure. On the other hand, peptide T4L143-155 was active under pressure against all bacteria except *E. coli* and *P. fluorescens*, two bacteria that were very sensitive to native lysozyme under pressure. HEL96-116 was more effective than T4L143-155 against all bacteria, and bactericidal activity of both peptides under pressure was not accompanied by spheroplast formation.

DISCUSSION

A first objective of this work was to investigate whether the previously reported observation that high pressure can sensitize *E. coli* to lysozyme can be extended to other gram-negative bacteria. It was found that this is the case for some but not for all gram-negative bacteria. Inactivation by lysozyme under pressure was concentration dependent since at 100 μg/ml more bacteria were sensitized and a higher magnitude of sensitization occurred than at 10 μg/ml. In line with what was reported earlier by Hauben et al. for *E. coli*, we found that sensitization is transient. As soon as pressure was released, all bacteria immediately regained their resistance to lysozyme. In addition to this type of transient sensitization, high pressure also causes a persistant type of sensitization, for instance to the lactoperoxidase system. Of course, whether or not an organism gets sensitized to lysozyme by high pressure may depend on many other factors, such as pressure, temperature, pH, medium composition, and cell growth stage and history. For example, we have previously demonstrated that application of pressure pulses with brief interruptions can cause sensitization to lysozyme and nisin of bacteria that are not sensitized by a continuous pressure treatment.

In the present work, we explored another route to maximize the synergistic bactericidal effect of pressure and lysozyme, by replacing native lysozyme with denatured forms of lysozyme and peptides derived from lysozyme, to both of which have been ascribed certain antimicrobial effects previously.

We prepared heat-denatured lysozyme according to Ibrahim et al. but we failed to confirm any of the bactericidal or bacteriostatic effects under atmospheric pressure that were reported by these authors. Denaturation with -mercapto-ethanol also did not endow lysozyme with antimicrobial activity at atmospheric pressure. We believe therefore that the effects described are very strain dependent.

The experiments with the denatured lysozymes under high pressure lead to two observations. One is that the enzymatic activity of lysozyme is required for it to exert a bactericidal effect under pressure. Reduction of enzymatic activity by heat or -mercapto-ethanol denaturation clearly leads to a reduction in the observed bactericidal effect, and complete elimination of enzymatic activity by extended heat treatment (60 min, 100°C) completely eliminates the bactericidal effect. At this point, therefore, our results do not allow us to confirm the hypothesis raised by other authors that the antimicrobial activity of lysozyme and/or heat-denatured lysozyme would consist partly of a mechanism that is independent of enzymatic activity. The second observation from these experiments is that partial heat denaturation can extend the spectrum of lysozyme bactericidal activity under pressure to a wider range of bacteria. In our experiments, heat denaturation made lysozyme active against serovar Typhimurium. A threshold level of residual enzymatic activity remains a requirement also in this case. The mildly denatured lysozyme H80/20-lys (with 11.5% residual activity) is active against serovar Typhimurium at 10 μg/ml under pressure, while the extensively denatured H100/20-lys (with only 0.6% residual activity) is active only at 100 μg/ml.

Taken together, these results allow us to formulate the following hypothesis about the synergistic effect of high pressure and lysozyme on the inactivation of gram-negative bacteria. At ambient pressure, lysozyme is completely inactive against most gram-negative bacteria because it cannot penetrate the outer membrane to reach its target, the peptidoglycan. Nevertheless, lysozyme has both cationic and lipophilic properties, which are known to contribute to an intimate interaction with and passage through bilayer membranes of many small peptides with antibacterial properties.

Passage is believed to occur through the so-called self-promoted uptake mechanism. Interestingly, it has been demonstrated that deep rough mutants of serovar Typhimurium are sensitive to lysozyme, suggesting that small changes in outer membrane composition may allow self-promoted uptake of lysozyme. Conversely, a change in the structure of lysozyme by heat denaturation can also cause the enzyme to become active against strains of *E. coli* which are insensitive to native lysozyme. Under high pressure, the ultimate mode of action of lysozyme remains the same, i.e., the peptidoglycanolytic activity. However, pressure apparently stimulates passage of lysozyme through the outer membrane of several gram-negative bacteria. We have previously named this phenomenon pressure-assisted self-promoted uptake, or, briefly, pressure-promoted uptake. Our present results show that pressure-promoted uptake of lysozyme is not a universal phenomenon in all gram-negative bacteria, probably because it depends on subtle properties of, and interactions between, the cell surface and the lysozyme molecule. Subtle changes in either the cell surface or lysozyme structure may change the outcome of the interaction. For example, an *E. coli* mutant has been decribed that is resistant to lysozyme under pressure.

In the present work, mild denaturation of lysozyme allowed successful pressure-promoted uptake and bactericidal action against serovar Typhimurium, which is refractory to pressure-promoted uptake of native lysozyme. The underlying explanation may be an increase in the hydrophobicity of lysozyme by heat denaturation, since it has been shown previously that increasing lysozyme hydrophobicity by genetic fusion of a hydrophobic peptide to the NH_2 terminal of the enzyme or by chemical modification of the lysyl residues with saturated fatty acids enhanced the activity of lysozyme against gram-negative bacteria. An alternative mode of antibacterial action that has been proposed for lysozyme is that the binding of lysozyme to the bacterial envelope would activate the autolysins. As a second part of this work, the bactericidal effect under pressure of two specific lysozyme-derived peptides to which antibacterial properties had been previously assigned was investigated. Again, we were unable to reproduce these effects with our panel of test bacteria and under our experimental conditions. However, both peptides were very effective under high pressure, in particular the peptide derived from hen egg white, which was active against all test bacteria. For these peptides, a different mechanism of bactericidal action must be involved because they are completely devoid of enzymatic activity. The precise mechanism remains unknown, but it seems evident that, similar to lysozyme, bactericidal activity will depend on passage through the outer membrane.

The use of lysozyme, heat-denatured lysozyme, and peptides derived from lysozyme may find interesting applications in the nonthermal preservation of foods and pharmaceutical and other products by high pressure. In the presence of these additives, lower pressures can be used to achieve the desired reduction factors, making high hydrostatic pressure technology more economically feasible. For optimal performance, the working spectrum of these compounds under pressure will have to be studied in more detail, and it may be necessary to design mixtures of specific compounds to cover as wide a range of bacteria as possible. For the peptides derived from lysozyme, toxicity studies should be conducted to demonstrate their safety before they can be applied in foods.

7

Microbiological Tests and Measurements

New chemicals are produced in increasing numbers. In Finland every year about 28 000 different products are manufactured or imported which can be classified as harmful. These products contain about 5000 different harmful substances. We also receive harmful compounds in airborne emissions. Substances are further transformed in industrial processes, in waste management and in the environment by human activities and natural processes. However, only rather limited monitoring data is available about the environmental concentrations of these compounds (concerning about 20–40 substances). The environmental risks of harmful substances can be recognized, assessed and managed by :

(1) measuring concentrations in the environment,

(2) testing the effects on biota in toxicity tests and

(3) monitoring ecosystem changes in the environment. Microbes have a key role in the environment as degraders in the carbon and nutrient cycles.

They also have the capability to transform and degrade many harmful man-made substances. Microbes have a strong effect on the exposure of other biota to chemical substances. Hence they can be regarded as primary targets for harmful effects of chemicals or active transformers of chemicals in the environment. The objective of this study was to evaluate the usefulness and applicability of microbial toxicity tests, biodegradation tests and microbial biomass and activity measurements in the environment for the environmental hazard identification and risk assessment of chemicals, effluents and contaminated soils. The study considers both aquatic and soil environments. The effects of modern pulp mill bleaching effluents on biota were assessed by a set of different biotests and by chemical analysis. The effluents and their long term effects on aquatic microbial processes were also studied in outdoor mesocosms. In one part the biodegradation kinetics of carbon-14-labeled model compounds in standard tests and in environmental conditions were measured and compared. The effects of several pesticides on soil microbes were tested and assessed in laboratory tests and in field trials. The applicability of microbial toxicity tests in the assessment of soil pollution and bioremediation processes was also studied in comparison with chemical analysis. The results of the study can support the development and selection of test methods for environmental risk management and regulatory decisions.

INTRODUCTION

Background

The increasing number and volume of various chemical substances used by industry, agriculture, households and other everyday activities poses a challenge to assess the environmental risks of this chemicalization. About 28 000 chemical preparations which can be classified as harmful to the environment are imported to or produced in Finland. These preparations contain 5000 harmful substances, many of which are used intentionally to satisfy the demands for production. The substances introduced in our society in various goods end up in the environment both during and after use. We also receive persistent compounds in global airborne emissions. These chemicals can cause environmental risks to human or other biota when they reach a critical concentration, which may be very low for certain substances.

Some harmful compounds are produced as byproducts or through environmental transformations, often as a result of microbial activity. Microbes can also enhance environmental pollution by transforming substances to more toxic, bio-accumulative and bio-available forms, for example by methylating mercury in lake sediments, water (Matilainen 1995) and soil (Verta et al. 1994, Rudd 1995) or by the reduction of arsenate As (V) to more mobile arsenite As (III) (Macur et al. 2001). Certain fungi might be potent producers of the atmospheric ozone-layer decreasing chloromethane (CH_3Cl) (Harper et al. 2001). However, microbes also have a key role in the environment as degraders of harmful compounds. Hence microbial populations have a remarkable effect on the exposure of other biota to chemical substances. Microbes can be either targets (soil fertility loss, food web malfunctioning) or active biodegraders and transformers when they are exposed to chemicals.

The environmental risks of harmful substances and pollution can be recognized, assessed and managed by measuring concentrations of the compounds in the environment, by testing the effects of chemicals and pollution on biota using toxicity tests, and by monitoring ecosystem changes in the environment (Fig 1). Chemicals management, risk assessment, labelling and possible restrictions are based on chemicals testing, on measured data about their existence in the environment and on possible recorded detrimental effects in ecosystems. There are continuing international efforts to harmonise these strategies globally at United Nations (UN) level, in the Organisation for Economic Co-operation and Development (OECD) and in a more structured way in the European Union (EU) member countries. OECD provides the test guidelines, strategies and principles for testing the various properties of chemicals (OECD 1993). These internationally developed, harmonised and validated test methods cover the testing of physical and chemical properties, effects on biotic systems, biodegradation and accumulation and human health effects. To rationalise the testing efforts and to avoid unnecessary testing, it has been agreed that if a chemical is tested in any OECD member country according to OECD guidelines in a laboratory with Good Laboratory Practice (GLP) quality assurance status, the test data should generally be accepted in all the member countries (OECD 1981a). This Mutual Acceptance of Data (MAD- principle) sets high demands for quality and ecological relevance to the test guidelines development.

The methods for environmental analysis are harmonised internationally for water, sediment, soil and waste in the International Standardization Organisation (ISO). They include chemical analysis, toxicity and biodegradation testing, microbiological quality analysis and other biological methods for ecological quality monitoring. The use of ISO standard methods and in EU member countries usually the corresponding European standards (EN) is sometimes mandated by EU directives (for example the Water Framework Directive 2000/60/EU) or recommended by international conventions such as the Oslo and Paris Convention for the Protection of the Marine Environment of the North East Atlantic (OSPAR) and the Helsinki Commission for Baltic marine environment protection (HELCOM).

Toxicity Testing within a Triad of Techniques

The use of toxicity tests as a complementary tool and not just a substitute for chemical analyses and biological surveys has been formalized in studies of aquatic sediments such as the Sediment Quality Triad (Chapman 1986). This strategy allows a combined "weight-of-evidence" interpretation of possible environmental impacts based on all three techniques (Sergy 1987). This could be applied equally and respectively to water, sediment and soil (Chapman 1996, Brohon et al. 2001).

Chemical analyses

Using chemical analyses we can analyze certain compounds in the environment or in the effluent which are the probable causes of toxicity. Chemical analysis can predict whether toxic or detrimental effects are likely to occur because of the compounds analyzed, but it cannot provide a proof other of the effects. Chemical analyses can seldom provide information about bioavailability, effects on biota, antagonism or cumulative effects of complex mixtures. They have been widely and traditionally used for regulatory purposes with specified limit values for substances, and they can also be used predictively to calculate expected environmental concentrations. Chemical analyses are also important for environmental monitoring of known harmful substances.

But do we know all the substances we should analyse for and are the analyses technically or economically feasible? The more complex our effluents are, the more probable is the failure to carry out analyses for the most important substances, if it is not even known what they are. It has been calculated that we might only have identified 10 % of the organic compounds in our surface waters (Galassi et al. 1992). The compounds are not stable and they are degraded and transformed by environmental conditions and microbes. The reference to "pure ignorance" about organic substances or more precisely "the ignorance of which we are not even aware"(Blumer 1975), is not without justification. Clearly chemical analyses must be supplemented with other kinds of information in environmental risk assessment.

Toxicity tests

Toxicity tests indicate whether a sample of chemical, waste, effluent, sediment or soil causes a toxic effect on biota. They can provide information about harmful effects of unknown substances.

The advantage of biological toxicity tests over chemical analysis is the direct assessment of the potential biotic impact without the need for extrapolation from the individual impacts of the different chemical constituents. Observed toxicity can be an early warning or even prediction of potential environmental damage. It also takes into account the effects of toxicant mixtures. Toxicity tests can be regulatory if a certain limit of toxicity is set, like the dilution based G- (Giftigkeit) or LID- (Lowest Ineffective Dilution) values in Germany (Steinhäuser 1995).

An important limitation of toxicity tests is that identification of the agent of risk is difficult.

Traditional biotests have been slow and laborious, but technical improvements (automation and scaling down of size) are making them more feasible for routine analysis. They have also been considered to lack the ecological relevance. Only a few species are tested, out of many present in the ecosystems. The other stressors (extreme temperature, water current, predators, disease) are not active in the laboratory tests. Simplistic use of the results might mislead the risk assessment.

However, the exposure of aquatic and soil biota to environmental chemicals, either as individual compounds or as complex effluents, may result in direct (killing, reproduction failure) or indirect effects (food web changes) which cannot be realistically assessed by even very complex model test systems with all possible target species. Often various single species screening tests are employed to detect the possible harmful effects of effluents on biotic systems. Although in the aquatic environment different species and complex food webs are exposed to effluents, the actual toxic impact is confined to generally a few relevant targets at the bio-molecular level. The functional units (respiration, enzyme functions, membrane transport) are often similar in different species. Hence single species tests and rapid in vitro screening tests could also be used to screen and classify the potential harmful effects of chemicals and effluents (Wenzel et al.1997). In order to simulate the aquatic food chain/web, representatives of different trophic levels (algae, daphnids and fish) are often used as tests organisms.

Biological surveys and environmental assessments

The biological survey of the ecological quality in a given environment is a definitive check on whether an ecosystem has been affected. Biological quality objectives and standards are being set in Europe in the implementation of the EU Water Framework Directive. A biological survey cannot serve in predictive mode (the harm is already done). Sometimes the responsible emission is difficult to identify and verify without toxicity data and chemical analysis of the possible toxicants even if all the other epidemiological principles are applied (Munkittrick and McCarty 1995). All three feet (concentration of the substance, proven toxicity and change in biota) are needed for the table to stand.

Basic Concepts of Ecotoxicology

In all environmental toxicity tests three aspects must be considered: the organism exposed, the medium for exposure (e.g. soil, sediment, water) and the measured characteristics (so-called "toxicity endpoint"). The endpoint is assessed or calculated by comparing the effects caused by the tested material to the effects found in control organisms in identical test conditions without the tested substance. In the most common test design the tested material is diluted in the test medium to give a series of concentrations.

The degree of toxicity of the test material can be estimated from the doseresponse curve, providing an estimate of the effective concentration (EC). The EC50-value, generally means the dilution or concentration of the tested substance, at which 50 % inhibition of the measured characteristics are observed compared to the unexposed control.

Depending on the test duration, conditions and organisms the tests can be considered as acute, sub-chronic or chronic. Chronic tests usually cover 10 % or more of the organism's life span (Sergy 1987). Death, growth, reproduction, enzyme activity, mobility or other activities are used as endpoints in the toxicity tests. There are test designs based on a single strain of one organism and also tests which apply to whole communities. The selection of a single strain or a community has a major effect on the sensitivity of the test, especially when using microorganisms. For example in the growth tests the community diversity changes can sometimes mask the toxicity which could possibly be observed by using only one sensitive microbial strain of at a time.

The bioavailability and actual concentration of the tested compound or material in the test systems is of crucial importance. The test medium (water, sediment or soil) and its characteristics determine the bioavailability and sometimes also the toxicity of the tested material. For example the hardness, buffering capacity and pH of water can have a strong effect on the toxicity of a substance. In the soil studies the organic carbon content or clay content can affect the bioavailability of many substances. In the chronic test with longer duration various, flow-through or semistatic test designs must be applied. The possible biodegradation or adsorption to the test vessel or to solid particles must also be considered and often the actual measured concentration of the tested substance can differ remarkably from the nominal concentration used in the test.

As with any analysis, quality assurance in the test design is important. It is most important to validate the test system every time with positive control such as reference toxicant and materials and also to use negative controls in order to be able to check the normal functioning of the test organism.

Microbiological Toxicity Tests

Here, microbiological toxicity tests are regarded as test designs using defined bacterial strains, mixed populations of environmental bacteria and to some extent unicellular algae. Bacteria and yeasts are also used widely as test organisms for genotoxicity and mutatoxicity testing. Some protozoan tests and tests using organelles (e.g. mitochondrial particles) resemble microbial tests but these tests are not reviewed here.

Various bacterial bioassays have been developed for toxicity screening of water, wastewater, sediment elutriates and soil extracts. In some recent tests solid samples (sediment, soil) can also be tested directly without any extraction. Three microbial tests have been developed and harmonized by OECD for testing the effects of chemicals, especially pesticides. Most of the tests are based on measurements of growth inhibition, respiration, mineralisation or metabolism, enzyme activity and viability of bacterial cells.

In the growth inhibition tests the cell multiplication is usually monitored by optical density measurements (ISO 10721, 1995 and ISO 15522, 1999) or by cell counting or fluorometry in algal tests (ISO 8692, 1986 and ISO 10253, 1998). The respiration tests are based on carbon dioxide production or oxygen consumption during the tests (ISO 8192, 1986). The mineralisation of added material containing organic nitrogen and production of metabolites is used for example in tests measuring nitrification (ISO 9509, 1989 and OECD TG 216, 2000). Several microbial enzyme activities can be measured directly by luminescence measurements (ISO 11348, 1998) or by specific substrate utilization and colour development (Reinharz et al. 1987). In one case the active movement of flagellated bacteria is used as an assessed characteristic (Bowdre and Krieg 1974). In anaerobic tests the production of gases is measured simply as pressure changes during the test (ISO 13641, 2001).

The cell walls of bacteria and other organelles differ to some extent from those of higher organisms (Bitton and Koopman 1992). Hence in some cases the ecological relevance of extrapolating the effects to other organisms might be limited. Especially the exposure routes and metabolism of harmful substances are sometimes unique to bacteria. The difference between gram positive and gram negative bacteria should also be considered.

The outer membrane of gram negative bacteria is an effective barrier to hydrophobic substances (Bitton and Koopman 1992). The permeability of the outer membrane can be enhanced by lyophilisation or by other means (Dutton et al. 1990)

Because of the small size, short multiplication times and easily measurable reactions the development of microbial tests is towards microscale and automated testing, on-line systems and field measurements (Blaise et al. 1998).

Tests for waters, effluents and aqueous solutions

Microorganisms, particularly bacteria, are attractive for use in wastewater toxicity testing because microbial tests are simple, rapid, sensitive and inexpensive (Bitton and Dutka 1986). By using bacterial toxicity tests it is also possible to estimate the toxic impacts of different, often fluctuating, untreated wastewaters on activated sludge wastewater treatment processes. Most modern wastewater treatments rely on microorganisms, and therefore their activity is of primary importance to the proper functioning of the treatment process.

Today the most common microbiological test in effluent toxicity assessments is the luminescent bacteria test (ISO 11348, 1998). Since the first studies on the use of luminescent bacteria for toxicity assessment of air pollutants by Serat et al. (1965), several test systems for routine use have been developed. The inhibition of light production by Vibrio fischeri bioluminescent bacteria indicates disturbance of the energy metabolism. The luminescent measurement assesses the metabolic status of this bacterium (Hastings 1978) and the change in bacterial bioluminescence after exposure to wastewater samples can be used as an indicator of potential toxicity (Bulich et al. 1981). A limitation of the use of marine bacteria such as V. fischeri, is the required high NaCl concentration (2 %) which is claimed to have effects on the bioavailability of certain compounds (Ribo and Kaiser 1987). One possible solution could be to use other solutes such as saccharose to balance the osmotic pressure during the test (Hinwood and McGormick 1987). Lampinen et al. (1990, 1992) developed genetically modified luminescent bacterium from Escherichia coli and Bacillus subtilis to overcome certain problems with natural luminescent bacteria used in toxicity testing. Luminescent bacteria test kits are also commercially available (for example Microtox TM, Microbics Co., Carlsbad, CA, USA; LumistoxTM, Dr. Lange GmbH, Berlin, Germany or BiotoxTM, Aboatox co., Turku, Finland).

Another standardized test for aqueous solutions is the Pseudomonas putida growth inhibition test (ISO 10712, 1995). The bacterium P. putida is a common aquatic heterotrophic microorganism. When cultured under specified conditions, in a defined medium with different concentrations of wastewater over several generations, toxic substances present in the wastewater sample can inhibit the cell multiplication of this bacterial species (Bringmann and Kühn 1977). Inhibition of the growth of activated sludge microorganisms is also used as a measured characteristic in a standardized test (ISO 15522, 1999). Growth inhibition tests of Pseudomonas fluorecence has been published by Paran et al. (1990).

In the activated sludge respiration inhibition test (ISO 8192, 1986), the oxygen consumption of the activated sludge supplemented with glucose is measured for four hours with an oxygen probe (King & Painter 1986). The nitrification potential of the activated sludge microorganisms can be assessed and inhibition of nitrate production from added ammonium salts can be measured (ISO 9509, 1989). The only standardized anaerobic microbial toxicity test (ISO 13641, 2001) is based on simple pressure measurements (anaerobic gas production) in sealed vessels.

The inhibition of enzyme activity or the inhibition of enzyme induction and de novo synthesis is assayed in some test procedures. The inhibition of Escherichia coli dehydrogenase activity is measured in a rapid ECHA screening test (Dutka and Gorrie 1989) and the inhibition of â-galactosidase activity in MetPAD (Bitton et al. 1992) and MetPLATE (Bitton et al. 1994). Inhibition of the hydrolysis of fluorescein diacetate (FDA) by several enzymes (proteases, lipases and esterases) and the measurement of fluorescence is applied in a test procedure presented by Haig and Rennie (1994). The inhibition of enzyme induction and de novo synthesis are considered to be more sensitive phenomena than the inhibition of enzyme activity (Dutton et al. 1988) or other metabolic activity in bacterial cells (Reinhartz et al. 1987). The Toxi- Chromo- test is based on inhibition of the activity of the lac-operon and of synthesis of â-galactosidase enzyme (Reinhartz et al.1987).

Inhibition of the induction of á-glucosidase in Bacillus licheneformis has also been reported to be especially sensitive to hydrophobic substances (Dutton et al. 1990). The greater sensitivity is caused by the difference of the cell wall of this gram-positive bacterium compared to E. coli.

The growth of unicellular micro-algae has commonly been used in toxicity testing of wastewaters, sediment elutriates and chemicals. These test designs can be regarded as microbial tests and the growth kinetics mainly follow the microbial exponential growth models. When exponentially growing cultures of green alga are exposed (for several generations) to various dilutions of a wastewater sample, the substances in the wastewater can either inhibit or stimulate the algal growth.

Standardized test protocols exist for effluent testing with the freshwater species Pseudokirchinella subcapitata (former Selenastrum capricornutum) in ISO standard 8692 (1989) and with marine species Sleletonema costatum and Phaeodactylum tricornutum in the standard ISO 10253 (1995). The inhibition of growth rate and biomass production are the measured characteristics. The tests are based on direct cell counts by automated or microscopic counting, or on measurements of chlorophyl production by spectrometry or direct fluorometry on microwell plates (Thellen et al. 1989, Mayer et al. 1997). Test designs using a set of algal species have also been published (Blanck et al. 1984).

Tests for solid samples

Often the toxicity assessment of solid samples (soil, waste or sediment) is based on the testing of aqueous elutriates or other extracts. Different solvents such as dimethyl sulfoxide (DMSO), methanol, ethanol and acetone can be used for extraction of organic contaminants (Kwan and Dutka 1990). One problem when using solvents is the possible toxicity of the solvent itself. Hence the maximum allowable concentration of the extract solvent in the test must first be measured (Brouwer et al. 1990) and the possible synergistic interactions between contaminants must also be considered (Gunkel et al. 1993). On the other hand aqueous extraction could underestimate the bioavailability of contaminants and hence their toxicity (Harkey et al. 1994). Aqueous extraction could be selective in multiple compound contamination in solid samples, due to the different solubilities of especially organic compounds in water (Ongley et al. 1988). The use of assays employing direct contact of test organisms with the sample could better assess the bioavailability and toxicity of the contaminants in solid samples (Rönnpagel et al. 1995). During the past decade several bacterial contact procedures have been developed to test the toxicities of soil, sediment and waste samples. The luminescent bacteria test could be performed either with soil water elutriates or with fresh homogenized soil samples with solid phase modification of the standard test (Brouwer et al. 1990). In this procedure the bacterial cells were exposed to solid samples and the solid particles were separated from the exposed bacteria by centrifugation before the luminescence measurement. The commercial Microtox TM Solid Phase (Microbics co., Carlsbad, CA, USA) used filtration as the separation technique of solids from exposed cells prior to luminescence measurement (Tung et al. 1990).

A novel possibility could be a kinetic modification of the luminescent bacteria test that could be used to assess soil or sediment samples (Lappalainen et al. 1999). In this method a kinetic measurement of the luminescence signal was started at the same time as the bacteria were added and mixed to the eluated sediment or soil sample. Luminescence was measured throughout the 30 s exposure and the possible luminescence inhibition was compared to the initial peak luminescence at the start of the test. This should also help to avoid some of the colour, turbidity or adsorption interference.

Kwan and Dutka (1992) applied the â-galactosidase enzyme synthesis inhibition test with E.coli (Toxi-ChromoPad test) to solid sediment samples. The direct contact application proved to be more sensitive than testing the sediment extracts.

The toxicities of soils and sediments could be tested directly and with high sensitivity with the procedure presented by Rönnpagel et al. (1995). They used the bacterium Bacillus cereus as test organism assessing the inhibition of dehydrogenase activity measured by rezasurin reduction (Liu and Strachan 1981). B. cereus is a common soil bacterium, has high affinity to soil particles and as a gram-positive organism its cell wall might be more permeable to hydrophobic organic compounds than the cell wall of the gram-negative bacteria used in most tests.

Testing of the effects of chemicals on microbes

The effects of chemicals and especially of pesticides on soil microorganisms have been studied for risk assessment because of the importance of microbes in soil processes. These studies mainly comprise laboratory studies such as respiration and mineralization inhibition tests. International harmonisation of these methods has only recently been undertaken by the OECD.

Harmful effects of a pesticide can often be observed in laboratory studies with soil types which favour high bioavailability of the chemicals. These 'worst case' studies are needed for the initial risk assessment of the pesticide. One such 'worst case' soil is for example the soil which was selected by an OECD working group for the assessment of effects of chemicals on soil microorganisms (70 % sand, and only 0.5–1.5 % organic carbon). Since in this soil the adsorption of the chemical is minimal and its bioavailability to microbes is maximal, tests with other soils are generally regarded as unnecessary. There are two harmonised OECD test guidelines for testing the effects of chemicals on soil microbes. Test guideline TG 217 (OECD, 2000) assesses the inhibition of the carbon mineralization of added organic material and TG 216 (OECD, 2000) assesses the inhibition of nitrogen mineralization of added nitrogen-containing organic material (lucerne meal). A more rapid method to screen the effects of chemicals on ammonium oxidation (the first step in the nitrification process) in soil based on chlorate inhibition of further reactions in nitrification was first published by Belser and Mays (1980), further developed for practical testing by Hansson et al. (1991) and proposed for harmonisation by Torstensson (1993).

In order to assess the effects of chemicals on activated sludge microorganisms and their respiration, a short term test TG 209 was developed by the OECD (OECD 1984). The OECD algae test guideline 201 (OECD 1984) is used for testing the effects of chemicals on freshwater algae. This guideline is currently under revision to improve the test procedure and to include more species from different taxonomic groups of algae.

Micro- and Mesocosms to Assess Effects on Microbiological Activity

Most aquatic ecotoxicological research on the assessment of effluents or chemicals is concentrated on single-species laboratory testing, or measurements of the effects on certain communities (fish, benthic fauna, algae) of the exposed ecosystems. Extrapolation from single laboratory toxicity tests to natural ecosystems is often uncertain. Conversely, studies on exposed natural ecosystems include much natural variation, both in terms of the sensitivity and exposure of the species and physicalchemical conditions, which complicates the interpretation of observed data. A possible compromise between single species laboratory tests and natural ecosystem tests could be a multi-species mesocosm that simulates the structural and functional mechanisms of natural ecosystems. These mesocosms are also affected by environmental factors other than the exposure studied, but often these factors can be recorded and hence the possible responses caused by the studied exposure could be better interpreted (Lehtinen et al. 1997). In mesocosm studies the ecosystem may develop certain characteristics over time (e.g. biomass increases and community production decreases (Odum 1985)). In effluent-exposed mesocosms or in stressed ecosystems these trends may be arrested, reversed or stimulated. Trends in stressed ecosystems include changes in community structure as well as its functions (nutrient cycling, energetics).

In soil studies harmful chemicals, such as pesticides, have effects on many parts of the food web (from microorganisms to soil invertebrates and plants). In order to understand the overall effects on the interactions between different organisms involved in mineralisation processes, terrestrial microcosm or mesocosm techniques can be used (Bruckner et al. 1995). These are usually assembled from different elements (microbes, invertebrates, plants) of the modeled ecosystem. They allow assessment of the effects of a particular variable on the system processes by eliminating or controlling that variable and monitoring the whole system responses (Ingham et al. 1986)

Microbiological Measurements in Environmental Pollution Assessment

Measurements of microbial activity or biomass can be used in the biological quality assessment of both aquatic and terrestrial environments. The possible observed changes are often a definitive indication of environmental damage. However, because of the microbial community changes in tolerance and species composition in the stressed environment, the measurable changes in total activity or biomass may be masked and are difficult to prove statistically (Giller et al. 1998).

The microbial activity assessments in environmental quality monitoring use turnover rates of certain substrates, enzyme activities and respiration as measured variables. The biomass estimates are much debated and these methods include measurements of ATP content of soil and sediments, soil carbon fumigation methods, different respiration measurements with or without added substrate and also microscopic observations of microbial cells. Developments in biochemistry and in molecular biology have improved the possibilities to assess changes in microbial community structure and functional diversity. These methods are reviewed separately for aquatic environments and soil.

Aquatic environments

Microorganisms of the sediment are important mineralizers of organic matter, which is the main source of energy in benthic food webs. This organic carbon and nutrient source consists of particulate organic matter and high molecular weight, polymeric compounds which first need to be enzymatically cleaved outside the microbial cells into smaller molecules before they are available for cellular uptake and further metabolism. Enzyme reactions by extracellular enzymes produced by algae, fungi and bacteria have been proposed to be "key steps" in aquatic food webs (Hoppe 1992).

Possible effects of effluents on the microbial loop of the aquatic food web can be assessed by measuring the microbial biomass or activity in the sediment and in the water column.

Aquatic microbial biomass can be measured by fluorescence microscopy counting with DAPI (4',6-diamidino-2-phenylindole) or acridine orange staining (Kepner and Pratt 1994), or by measuring the extractable ATP content of the sample (Karl and LaRock 1975).

Microbial activity is often assessed by measuring enzyme activities (phosphatases, ureases, dehydrogenases) or by measuring the heterotrophic activity for example by methods based on incorporation of radiolabelled thymidine (Riemann and Lindgaard- Jörgensen 1990 or leucine (Chin-Leo and Kirchman 1988). Field studies have shown that extracellular enzyme activity measurements can be used in the biological quality monitoring of sediments (Lee and Tay 1998).

Activity measurements of phosphomonoesterase (PME) can be used as an index of potential organic phosphate mineralization. The activity of this aquatic extracellular enzyme may be potentially increased by eutrophicating effluents, e.g. from sewage (Chappel and Goulder 1994) or fish farms (Carr and Goulder 1990, Davis and Goulder 1993), or inhibited by toxic effluents (Burton and Lanza 1987, Tubbing and Admiraal 1991).

The functional diversity of the sediment microbial population and its extracellular enzyme activities can also be assessed by enzyme activity profiles (Poremba 1994).

Terrestrial environments

In field studies or in monitoring soil fertility and in assessing the effects of anthropogenic activities on microbial populations, measurements of microbial activities related to carbon and nitrogen metabolism are used. These measurements, such as soil respiration (Nordgren 1988) and nitrification (Müller et al. 1981), were adopted in the EPPO/ Coe pesticide risk assessment guidelines (OEPP/ EPPO 1994) to assess long term effects of pesticides on soil fertility.

The generally most accepted estimates of soil microbial biomass are considered to be obtained by the soil fumigation incubation method (FI) (Anderson and Domsch 1978), the fumigation extraction method (FE) (Vance et al. 1987) or by the substrate-induced respiration method (SIR) (West et al. 1986). Microbial active biomass can also be assessed by measuring the amount of active microbes by ATP measurements (Jenkinson et al. 1979, West et al. 1986, Arnebrant and Bååth 1991,Vanhala and Ahtiainen 1994).

Fluorescein diacetate (FDA) hydrosis by many enzymes (non-specific esterases, proteases and lipases) has also been used to determine the total microbial activity in soils (Schnürer and Rosswall 1982, Adam and Duncan 2001). The FDA hydrolysis measures the activities of both membranebound enzymes and free extracellular enzymes (Stubberfield and Shaw 1990). Several other individual enzyme activities, for example urease (Tabatai 1977), phosphatase (Tabatai and Bremner 1969), dehydrogenase (Aoyama and Nagumo 1995) and glucosidase (Barnhart and Vestal 1983) activity have been proposed as tools to assess soil microbial activity.

The determination of bacterial growth rates in soil by measuring incorporation of tritium-labelled thymidine (Bååth 1992) or ^{14}C-labelled acetate to lipids (Barnhart and Vestal 1983) has also been published as estimates of microbial activity in soils.

The above-mentioned sum parameters of microbial activity can indicate a severe drop in total microbial community, but may fail to indicate possible harmful changes in microbial community structure. The effects of a pesticide may be very different on fungi and bacteria (Jones et al. 1992, Malkomes 1992).

The diversity of microbial consortia in soil can be assessed by methods based on nucleic acids (Torsvik et al. 1990, Führ and Kubiak 1994,) and phospholipid fatty acid profiles (Bååth et al. 1992), and the functional diversity can be described by substrate utilisation patterns (Degens and Harris 1997) or by soil enzyme activity profiles (Kandeler and Eder 1993, Bandick and Dick 1999, Vepsäläinen et al. 2001). Nannipieri et al. (1990) showed that the assessment of enzyme profile could be a better estimate of overall microbial activity than a single enzyme measurement when evaluating responses to pollution.

As a complex matrix the soil sets high demands on the applicability of study methods of any biological activity and also on the interpretation of the results. In field studies, natural soil with clay and a high organic carbon content may have such a high adsorption capacity that no effects of chemicals on biota are detected. In Finland, most of the cereal crop production and the use of pesticides is located in areas with high clay and also organic carbon content of soil. This may have effects on the bioavailabilities of the pesticides used.

The methods for soil biological quality assessments are standardized by the ISO Soil Quality Technical Committee 190. Currently there are only two accepted methods to assess soil microbial activity and biomass. The determination of nitrogen mineralisation and nitrification in soils (ISO 14238, 1997) is based on the measurement of ammonification and nitrification after addition of organic nitrogenous matter. The other method is for the determination of soil microbial biomass by fumigation and extraction (ISO 14240-2, 1997). A rapid screening method for soil nitrification based on ammonium oxidation has been proposed as a soil quality standard by Torsstensson (1993).

Biodegradation Studies on Chemicals

The natural carbon cycle in ecosystems operates on the assumption that practically all biosynthetic organic materials are biodegradable. Natural compounds of low degradability and man-made compounds may ultimately be degraded but this may require some time for adaptation. The potential for this adaptation and for radical population changes exists in microbial communities. However, the environment will often be ill prepared to withstand vast and increasing amounts of compounds from human activities.

The estimation of biodegradation rates is an important uncertainty factor in the chemical risk assessment mandated by EU legislation (European Commission 1999). Obviously the rate of biodegradation greatly influences the predicted environmental concentrations (PEC) and thus the possible exposure of biota to the chemical.

The existing OECD tests for ready biodegradability have been developed to devise screening methods to determine whether a chemical has the potential to be easily biodegraded, and not to predict the rate of biodegradation in the environment. In these tests the potential biodegradability is classified usually by evaluating the mineralization of the chemical by measuring the decrease of organic carbon, carbon dioxide evolution or oxygen demand during the test. In some tests only the primary biodegradation is assessed by analyzing the disappearance of the original parent compound tested (not mineralisation). In the OECD regime a compound is regarded as readily biodegradable if it reaches 60–70 % degradation during the 10 day period from the start of the degradation (OECD 1981). However, current risk assessment and future EU legislation require degradation rates. Usually these rates must be judged by experts (default values in the EU technical guidance document for the risk assessment of new and existing substances, EC 1999) or derived by modelling e.g. by the Simple Treat fate model (Struijs et al. 1991) for wastewater treatment plants based on the available data from ready biodegradability tests.

The fate, biodegradation and chemical transformation of chemicals such as pesticides in soil environments is usually studied in soil columns and with labeled substances. For testing the aerobic and anaerobic transformation and degradation of chemicals in soil, OECD is currently finalizing the guidelines. Hitherto OECD has been able to harmonize only one guideline, TG 304A for inherent biodegradability in soil (OECD 1984).

The several ISO Water Quality standards for testing the biodegradability of organic chemicals or any organic material (effluents) provide strict technical guidance for testing and for test validity. They do not contain guidance for interpreting the results as substances being "readily biodegradable" as do the OECD guidelines with set criteria. The two existing ISO Soil Quality standards for testing aerobic (ISO 11266, 1994) and anaerobic (ISO 14239, 1997) biodegradation of chemicals in soil provide guidance to perform the testing. These ISO standards are to be considered as methods for assessing the capability of soil to degrade organic chemicals rather than for chemicals testing.

Biodegradability depends not only on the molecular structure of the test compound but also on the microorganisms available and on other environmental conditions. The feature of adaptation of the microbial population in the tests, in the wastewater treatment and in the environment is a key issue in the biodegradability testing and in the interpretation of biodegradability data. The standard tests (OECD Test Guidelines and ISO standards) have many arbitrary features compared to actual environments, in order to improve practicability. An important difference is the concentration of the tested chemical in the standard tests and in the aquatic environment or wastewater treatment plants. The disadvantage of the summary parameters measured in the standard tests (decrease of organic carbon, carbon dioxide evolution, oxygen demand) is that higher concentrations of the tested compounds are required than presumably occur in different environmental compartments.

Concentrations of the chemicals in the treatment plants or in the environment are usually at ng–mg carbon per litre levels compared to 20–400 mg C/L in most tests. The different concentrations of the chemical/substrate will provide different growth kinetics and hence different biodegradation rates (Ingerslev et al. 1998). This correlation between test compound concentrations and biodegradation rates has been reported for many substances, revealing a decrease in rates at lower concentrations (Boethling and Alexander 1979, Alexander 1994).

At high concentrations the chemical, if it is biodegradable, can be degraded as a primary substrate and the competent microorganisms grow exponentially, resulting in an accelerating biodegradation rate (Monod kinetics). At low concentrations in the environment, with natural carbon substrates which are degraded simultaneously, the chemical cannot serve as a primary substrate and thus will not support significant growth of the degraders and the biodegradation follows first order kinetics with a constant biodegradation rate. However, adaptation of the microbial community may eventually take place, mediated by constant exposure to the chemical. This may lead to faster degradation, and hence the rates in different environments (adapted/ non-adapted) may differ considerably.

Objectives of this Study T

oxicity tests based on the responses of different organisms have been used to assess the environmental hazard identification and risk assessment of chemicals, waste waters, contaminated sediments and soils for some decades. There also exist methods to evaluate changes in the ecological quality of the various environments based on the structural or functional changes of the biota. Microbiological methods have been applied among other measurements to achieve these goals.

The aim of this study was to evaluate the applicability of different microbial toxicity tests (Papers I–V) and microbiological activity or biomass measurements in the environment (Papers II, V) for the effect assessment of chemicals (Paper V), waste waters (Papers I and II) and soil pollution (Papers III and IV). The usefulness of these microbial methods for the regulation and guidance human economic activities (industry, wastewater treatment, agriculture, soil remediation) was also discussed. The sensitivity of assessment of risks or environmental changes was also compared with tests using other organisms and biological quality measurements and chemical analyses (Papers I–V).

The biodegradability of a compound is one of the most important factors affecting the exposure of biota in the environment to this compound. Hence the biodegradation tests are used to provide an indication of the fate of the substance in the wastewater treatment or in the environment. The aim of the study in paper VI was to compare the biodegradation of two reference compounds in a standard method and in more ecologically relevant conditions in a simulation test with different sources of microbial inocula.

MATERIALS AND METHODS

Study Structure and Methods

Assessment of environmental effects of industrial effluents

In the first phase of the study the toxicity of the wastewaters originating from conventional chlorine (Cl2) bleaching, novel elementary chlorine free (ECF) and totally chlorine free (TCF) bleaching processes was assessed and compared with their chemical composition. In the second part of the work, more detailed data concerning the acute toxicity of the different wastewater streams of modern Finnish pulp mills were collected in order to identify the sources of toxicity in different wastewater streams in the mills and to assess the toxicity reduction efficiency of secondary treatment. In the third part of the study (previously unpublished data) the toxicity and the toxic loads from Finnish pulp and paper mills were compared to those from some other industrial areas such as the metal and mining, chemical and textile industries.

The two pulp mills chosen for the first phase of the study produced bleached hardwood and softwood kraft pulp (Table 5, Verta et al.,1996). One of the mills (Mill A1) produced non-bleached board and bleached and non-bleached kraft pulp and the other (Mill B1) produced bleached kraft pulp on two separate hardwood and softwood lines. Both the bleaching waste waters (1 chlorine bleaching, 4 ECF bleaching and 3 TCF bleaching) and the secondary treated effluents (1 chlorine bleaching, 3 ECF bleaching and 3 TCF bleaching) from these mills were sampled and studied. In order to obtain an understanding of the toxic effects, a set of acute and sub-acute bioassays was applied (Pseudomonas putida growth inhibition, Vibrio fischeri bioluminescence, Pseudokirchinella subcapitata algae growth inhibition, Daphnia magna mobility inhibition (ISO 6341, 1989), Brachydanio rerio zebra fish embryo and sac-fry stage test on fertilized eggs and hatched larvae (SFS 5501, 1991), see 2.2.1 and Ahtiainen et al., 1996 for details)

In the second phase of the study nine pulp and paper mills, practically all the significant bleached pulp producers in Finland, were chosen for the study (Table 6, Ahtiainen et al., 2002). They produced both bleached hardwood and softwood pulp. The process used was either ECF-, TCF- or combined ECF/TCF-bleaching. The untreated wastewaters were collected from debarking (n = 7), black liquor condensates (n = 4), different bleaching sequences (alkaline (n = 19), acidic (n = 19) and total bleaching waste water (n = 18)), total untreated wastewater (n

= 15) of the mills and secondary treated (activated sludge) effluents (n = 10) as pooled samples representing the normal functioning of the factory processes. The acute toxicity and the influence of freezing and storage of the samples on the toxicity of the wastewaters were assessed by 2 simple screening tests (luminescence inhibition test and the mitochondrial particle test, see 2.2.1. and Ahtiainen et al., 2002 for details. In the third part of the study, in which the toxicities (P. putida growth inhibition, V. fischeri bioluminescence, P. subcapitata algae growth inhibition, D. magna mobility inhibition) of various industrial effluents were assessed, the pulp- andpaper industry was represented by a bleach pulp mill, a non-bleach pulp mill, a non-bleach cardboard mill and a paper mill.

Chemical industries were: oil refinery and petrochemical plant, pharmaceuticals production, biotechnical enzyme production and viscose production. Metal industry was represented by stainless steel production, cobalt and zinc production and mining. Textile factories produced leather and yarn. The two effluent samples were collected at each mill as pooled composite samples of one day.

The general chemical characteristics of the effluents; biological oxygen demand (BOD7), chemical oxygen demand (CODCr), total organic carbon content (TOC), absorbable organic halogens (AOX), total nitrogen (Ntot), total phosphorus (Ptot), loss on ignition, suspended solids and colour, were analysed using standardized methods (Verta et al. 1996). In the first phase of the study 25 major and trace elements, phenolic compounds, fatty acids, resin acids and molecular weight (MW) distribution of the wastewaters were also determined (Verta et al. 1996).

Effects on aquatic microbiological processes in mesocosms (Paper II)

The long term effects of untreated and treated ECF and TCF pulp bleaching wastewaters, secondary treated municipal effluent and natural humic water on microbial processes (sediment enzyme activities, see 2.2.3) and on biomass (water column bacterial counts and sediment ATP content, see 2.2.3) were studied in brackish water mesocosms. The effects on other parts of the aquatic food web were also studied (Tana et al. 2002) in this one and a half year experiment. For additional background information the toxicities of all samples were assessed by a set of biotests: luminescent bacteria test, algae test and daphnid test. The tested water, wastewater and effluent samples were also characterized by different chemical analysis (Mikkelson and Paasivirta 2002) and the correlation between analysed compounds and mesocosm effects was assessed (Paasivirta and Mikkelson 2002).

The mesocosm ecosystem included macro- and micro algae, invertebrates and juvenile sticklebacks (Gasterosteus aculeatus) indigenous to the Baltic sea area. The detailed mesocosm experimental setup is described by Tana et al. (2002) and in paper II. The untreated and secondary treated ECF and TCF effluents were studied at two concentrations, and the treated municipal effluent and humus water at the higher concentration. Hence the experiment consisted of two control pools and 10 single mesocosms receiving test waters. All waste water and treated effluent samples were collected as pooled samples and the humus water as a single homogenized sample. The samples were frozen immediately after sampling and stored in a freezer (–18°C) until thawed and diluted to the flow-through system of the mesocosms. The two dilutions, prepared in natural brackish seawater of the tested sample waters, were relatively low (1/ 400 and 1/2000) and were designed to represent the concentrations in rather large areas in pulp mill effluent recipients in Finland.

Toxicity assessment and bioremediation of polluted soils

Chlorophenol contaminated soil (Paper III) : The degradation of chlorophenols and possible formation of harmful metabolites was studied by chemical analysis and toxicity testing during the pilot-scale composting (6 months) of chlorophenol contaminated sawmill soil and impregnated wood. Four 13 m^3 compost piles were constructed, with nutrient addition, bark chips and pH adjustment by granular lime in the field at the end of May, at Vääksy, Southern Finland (Laine and Jörgensen 1997). One pile was left uninoculated as control, the second was supplemented with straw compost as inoculum and the third and fourth with remediated soil as inoculum.

The composite soil samples were homogenized and sieved through an 8 mm sieve to remove the added bark chips and other coarse particles before the toxicity assessments and chemical analyses (chlorophenols, extractable halogen content (EOX), molecular weight distribution of organic halogen compounds and dioxins and furans, see Paper III).

Creosote oil contaminated soil (Paper IV)

The bioremediation of creosote contaminated soils was studied in two cases of pilot outdoor composting by toxicity assessments and chemical analyses. In both cases the soil from the sawmill area was heavily contaminated with polyaromatic (PAHs) and aromatic (AHCs) hydrocarbons and with metals (As, Cr_2, Cu) from wood preserving activities.

At one site, two pilot composts (5m3) were constructed in the field near Vilppula in southern Finland. One of the piles was inoculated with Mycobacterium inoculant and the other was left uninoculated, thus containing only the "inherent inoculum" of indigenous bacteria. The composts were sampled at the start, after the first summer and after the second summer for chemical analyses and toxicity tests. The pooled soil samples were homogenized and sieved through an 8 mm sieve to remove the added bark chips and other coarse particles.

The other site was located at Ilmajoki, in western Finland. A large composting pile (100 m^3) of soil from an old wood preserving facility was constructed in the field. The soil was screened to remove >50 mm diameter stones and other coarse materials, and then pretreated with 50 % hydrogen peroxide to speed up the breakdown of the recalcitrant, but biodegradable 4–6 ring PAH compounds. After the hydrogen peroxide had completely degraded (tested by a fieldstrip test), a microbial inoculum of PAH degraders, nutrients (N, P and micronutrients) and bark chips was added. The pile was sampled for toxicity tests and chemical analysis of PAH composition at the beginning, after the first week, and then every four weeks during a composting period of 163 days. The compost was thoroughly mixed before sampling and the pooled soil samples (during one summer) were pretreated for analysis as in the first case.

Effects of agrochemicals on soil microbes (Paper V)

The study was designed to assess the effects of two pesticide regimes (conventional full insurance and supervised/reduced pesticide use) and two cultivation techniques (conventional tillage and fertilizing compared to no tillage with less fertilizing) on soil microbes in a field study, and to evaluate these results with laboratory tests. The herbicides used were chlorsulfuron, methyl-chlorophenoxyacetic acid (MCPA) and bentazone; the fungicides carboxin-imazalin and propiconazole; and the insecticides dimethoate and pirimicarb.

Laboratory Experiments

The potential harmful effects of pesticides (commercial formulations) on microbes were studied in the laboratory with two bacterial toxicity tests (P. putida growth inhibition and V. fischeri luminescence tests) in aqueous solutions. The soil respiration inhibition potential was measured for five selected pesticides and the bioavailability of three pesticides in the soil was assessed by a solid phase modification of the luminescent bacteria test.

Soil basal respiration inhibition by dimethoate, propiconazole, chlorsulfuron, pirimicarb and MCPA- bentazone was measured using an automated respirometer (Nordgren 1988). The same clay soil (clay >70 %, organic carbon content 1.5– 2 %) as in the field studies was supplemented with different amounts of pesticides (10, 50, 100, 500 mg/kg as commercial formulations) and the respiration was followed for two months.

To assess the bioavailability of the pesticides in the soil respiration inhibition tests a solid phase modification of the luminescent bacteria tests was used. The toxicity of each soil sample supplemented with different amounts of dimethoate, chlorsulfuron and propiconazole (10, 50, 100, 500 mg/kg as commercial formulations) was assessed by the soil-contact modification of the standard luminescent bacteria test.

Field Experiments

The effects of chemicals on microbial biota were assessed in the field studies by measuring microbial activities related to carbon and nitrogen metabolism (soil respiration, nitrification potential) and biomass (soil ATP content).

The field studies were carried out near Jokioinen in southern Finland. The size of the six replicate study fields was 54 m x 120 m. There was some slight heterogeneity between the randomized study fields set in the area in organic matter and clay content, but all the soil in the area would be classified as clay soil. The soil had not been treated with any pesticide during the seven preceding years.

During two growing seasons (1992 and 1993) of barley, different herbicides, fungicides and insecticides were used according to the general guidance calendar (treatment C) or to the actual monitored need for plant protection (treatment D). The herbicides used were chlorsulfuron, MCPA, bentazone, the fungicides were carboxin-imazalin and propiconazole and the insecticides were dimethoate and pirimicarb. In addition, the effects of two different cultivation techniques were studied. These were a normal tillage and fertilizer use (treatment A) and cultivation without tillage and with reduced use of fertilizers (treatment B). Hence the four different treatment compilations were: AC conventional pesticide use, normal tillage; AD reduced pesticide use, normal tillage; BC conventional pesticide use, no tillage, less fertilizers; BD reduced pesticide use, no tillage, less fertilizers.

The sampling was performed before and after pesticide treatments, and the soil was also sampled twice at the end of the growing season in order to assess the longer term effects. From each field, 30 surface soil (0–5 cm) core samples were pooled together. These sieved samples were kept at +4°C for a maximun of four days before analyses of nitrification potential and ATP content. Samples intended for soil respiration measurements were frozen immediately at –20°C.

Biodegradation studies (Paper VI)

Biodegradation of the reference chemicals aniline and 4-chloroaniline was measured and compared in a standard method at standard concentrations (10–20 mg carbon per litre) and in more environmentally realistic conditions at low concentrations (2–26 mg carbon per litre) with ^{14}C- labelled substances and with different sources of inocula. In the standard test the tested substance was added to a mineral medium as the sole source of carbon, and the sealed vessels with a headspace of air were inoculated with diluted fresh activated sludge from a municipal wastewater treatment plant in Helsinki.

At low concentrations of labelled reference chemicals in the shake batch test the test waters (inocula) were collected from two pristine lakes in southern Finland and from unpolluted Baltic seawater and urban seawater. One of the pristine lakes, Lake Vitsjö in Tenhola, was oligotrophic and the other, Lake Matkunlammi in Hyvinkää was a naturally humic lake with high carbon content. The unpolluted seawater was collected near Tvärminne on the southern coast of Finland and the urban seawater from Helsinki harbour. The waters were stored cool (4°C) and used for the tests on the day after collection. The numbers of heterotrophic bacteria in the natural test waters were enumerated by total aerobic viable counts at 20°C on R2A agar incubated for 3 days. The total organic carbon (TOC) content and some other water quality characteristics (pH, O_2, Ntot, Ptot) were also analysed.

The biodegradation rates of labelled reference chemicals were also assessed in the standard test conditions (suspended solids 4 mg/l) with activated sludge from the Helsinki municipal wastewater treatment plant. Biomass evolution during the tests was followed by ATP measurements and on the basis of residual ^{14}C- activity in the particulate matter.

Analytical Methods

Microbial toxicity test methods on effluents, soil elutriates and chemicals

Vibrio fischeri luminescent bacteria test : The toxicities of waste waters (Papers I and II), soil elutriates (Papers III,IV and V) and chemicals (Paper VI) were tested according to the standardized luminescent bacteria test (ISO 11348, 1998). In order to test soil elutriates, freshly homogenized soil samples were eluted to deionized water by weighing 5 g of soil to 50 ml of deionized water and shaking them overnight at room temperature. The pH of the elutriates and waste water samples was usually neutralized to pH 7 to give comparable results.

The luminescent inhibition test was accomplished by combining a 500 µl sample with 500 µl luminescent bacterium Vibrio fischeri DSM 7151 (same as NRRL B-11177) suspension. The test tubes were then incubated at 15°C in a water bath. Luminescence after 30 min incubation was measured with a luminometer (BioOrbit 1253). The inhibition of luminescence

(%) was calculated by comparing sample luminescence to that of the control containing deionized water.

Pseudomonas putida growth inhibition test

Effluent samples (Paper I), soil elutriates (Paper IV) and pure chemicals (pesticides, Paper VI) were tested by automated modification of the standard Pseudomonas putida growth inhibition test (ISO 10712, 1995) with the Bioscreen C^R analyzer (Labsystems). In this test, P. putida MIGULA (DSM 50026) bacteria were grown at 21°C for 16 hours in a liquid medium (350 µl) in special multiwell plates and the turbidity due to bacterial growth was measured by vertical photometry. The toxicity was assessed as growth inhibition (%) in different dilutions of the samples compared to a deionized water control. The pH of the samples was checked and when needed set at 7. The waste water samples were pretreated by centrifuging (10 min/10 000 rpm) to eliminate coarse particles and by sterile filtration (0.2 µm) to eliminate background bacterial growth.

Pseudokirchinella subcapitata green algae growth inhibition test

The algal growth inhibition testing of the sterile filtered effluent samples (Papers I and II) was performed by the ISO standard method (ISO 8692, 1989) with the alga Pseudokirchinella subcapitata (formerly Selenastrum capricornutum). Algal growth during 72 h was measured every day without chlorophyl extraction and analysis using whole cell fluorometry (Sequoia-Turner 450 Digital Fluorometer, filters: NB440 (excitation) and SC665 (emission)). The method was calibrated by microscopic counting of algal cells. The toxicity was assessed as growth inhibition (%) in different dilutions of the samples compared with deionized water.

Toxicity test methods on soil

Soil contact test method with luminescent bacteria : To assess the toxicity of soil samples (Papers III and IV) and bioavailability of the pesticides (Paper V) a solid phase modification of the luminescent bacteria tests was used (Brouwer et al. 1990). Homogenized soil samples of 5 g were weighed to 50 ml centrifuge tubes and diluted with 16 ml deionised water. This suspension was supplemented with 2 ml 20 % NaCl solution and with a 2 ml inoculum of Vibrio fischeri (strain DSM 7151) luminescent bacteria. The tubes were incubated for 15 minutes at 15°C in a water bath. After the incubation the tubes were quickly centrifuged (5 min, 1660 rpm, 300 g) to separate the solids from the exposed bacteria. The luminescence of a 1 ml sample of this supernatant was measured with a luminometer (BioOrbit 1253). The sample luminescence was then compared to that of deionized water to calculate inhibition percentage and to estimate EC50 values (50 % inhibition of luminescence).

Toxi-Chromo Pad enzyme synthesis inhibition test

The qualitative or semi-quantitative Toxi-Chromo Pad test (Kwan 1995) with soil samples (Paper IV) was carried out according to the instructions of the kit. Freeze dried Escherichia coli bacteria were rehydrated and added to the reaction mixture. After 20 min pre-incubation 0.5 ml of this solution was mixed with 0.5 g of homogenized fresh soil. The mixture was incubated at 37°C for two hours. The mixture was homogenized again and a small drop of incubated soil-bacteria mixture was pipeted onto a chromogenic pad (a glass microfibre filter soaked with blue chromogenic substrate for â-galactosidase). The pads were further incubated at 37°C overnight for the colour development. The intensity of enzyme synthesis was indicated by blue colour and possible toxicity by reduced the enzyme synthesis and weaker colour development.

Soil respiration inhibition test

Soil respiration inhibition of pesticides (Paper VI) was measured by automated respirometry (Nordgren 1988). The same clay soil as in the field studies was supplemented with different amounts of pesticides. The soil respiration was measured on several two-day periods during one month of incubation at 20°C. The soil moisture content was adjusted to 50 % of the water holding capacity in the beginning and after each week. The mean respiration rate in treatments was then compared to that of non-treated similar control soil.

Methods for microbiological activity and biomass measurements

Phosphomonoesterase activity in sediment **:** The alkaline phosphomonoesterase (PME) activity of the sediment (Paper II) was measured by the modification of soil phosphatase activity method based on a method in which phosphate is released from p-nitrophenyl phosphate (PNP-P) (Tabatai and Bremner 1969). Samples of 2 g fresh, homogenized surface sediment were incubated for 2 h with p-nitrophenyl phosphate (PNP-P) at pH 7.75 (0.1M Tris + 2mM EDTA buffer) at room temperature. The concentrations of p-nitrophenyl (PNP) were measured spectrophotometrically and the phosphatase activity was calculated from the PNP concentrations compared to blank samples.

Sediment enzyme activity profile with the Api Zym- kit

The profile pattern of enzyme activity diversity in mesocosm sediments (Paper II) was assessed semi-quantitatively by the commercial API Zym (BioMerieux, France) enzyme activity kit (Poremba 1994.). A fresh sediment sample of 2 g was homogenized with 8 ml of sterile deionized water. The tubes were centrifuged for 5 min to remove the solid particles. From this supernatant a drop of 65 μl was pipetted to each well of the API Zym strips. The strips were incubated for 16 h at 20°C. The enzyme activities were recorded according to the manufacturer's instructions and expressed as nmol activity per well during incubation time.

Epifluorescence microscopic enumeration of aquatic bacteria

The microbial biomass of the water columns in mesocosms (Paper II) was estimated by DAPI (4',6-diamidino-2-phenylindole) staining and epifluorescence microscopy counting (Leitz Diaplan microscope) (Kepner and Pratt 1994). Water samples of 50 ml were fixed by 5 ml of sterile filtered 10 % glutaraldehyde fixative immediately after sampling. For staining, 4.5 ml of sample and 0.5 ml of dispergant (44.6 g $Na_4P_2O_7$ x 10H2O) were homogenized for 30 min in an ultrasound homogenizer. Subsamples of 1 ml were stained with 20 µl DAPI (10mg µl-1) solution in 980 ml deionised water for 7 min and filtered onto black 0.2 µm pore polycarbonate filters (Nuclepore Costar 110656). The visible fluorescent cells were counted from at least 30 fields on filters. The biomass was expressed as viable cells per 1 ml of sample.

Adenosine 5'-triphosphate content in sediment, soil and during biodegradation tests

The active microbial biomass in sediment (Paper II), soil (Paper V) and during biodegradation tests (Paper VI) was assessed by measuring the ATP (adenosine triphosphate) content (Vanhala and Ahtiainen 1994). To extract ATP, samples of 10 g of fresh and homogenized sediment (two replicates, Paper II) or sieved soil (six replicates, Paper V) were placed in 100 ml vials containing 10 ml 20 % trichloroacetic acid and 10 ml 8 mM EDTA.

Suspended solids were removed by paper filtration and 0.5 ml aliquots of these filtrates were mixed with 0.1M Tris + 2 mM EDTA buffer in Eppendorf tubes. For the measurement the samples were further diluted 50 times with the same buffer. The ATP content was measured with a BioOrbit luminometric assay kit in a BioOrbit 1251 luminometer (BioOrbit, Turku, Finland). The ATP contents of the samples were then calculated on the basis of the fresh weight of the samples.

To assess the biomass evolution during the biodegradation tests a sub-sample of 0.5 ml was taken from the bottles during each sampling and frozen to –20°C for the ATP analysis. The ATP content of the sample was analyzed according to the protocol of Bio Orbit ATP Kit (BioOrbit, Turku, Finland).

Nitrification potential of soil

Soil nitrification potential (Paper V) was assessed as the capacity of the soil to oxidize ammonium ions to nitrate ions (Müller et al. 1981). The moisture contents of six replicate 50 g (dry weight) fresh soil samples were adjusted to 50 % water holding capacity (WHC) and 50 ml of 20 mM (NH4)2SO4 solution was added to the soil in 250 ml Erlenmeyer flasks. The flasks were incubated at 20°C for 20 days and the concentration of NO3 - ions was measured using a nitrate electrode (Orion 920 pH-meter). The nitrification potential (NO3-N mg g-1) was calculated per dry weight.

Soil basal respiration

The basal soil respiration rate (Paper VI) was measured by weighing six replicate 60 g (dry weight) samples of soil in respirometer cuvettes. The soil moisture content was adjusted to 50 % of the water holding capacity (WHC) and samples were stored for one week at 20°C to stabilize the respiration level. The cuvettes were placed in the respirometer, which captured CO2 in KOH and measured the change in conductivity (Nordgren 1988). The soil respiration was measured for one week at 20°C.

Biodegradation test methods

Standard headspace CO_2 evolution test : The biodegradation was tested according to the ISO 14593 headspace CO2 evolution test (1999) at standard concentrations (20 mgC l-1) of the test substance. In this test the tested substance was added to a mineral medium as the sole source of carbon, and the three replicate sealed vessels with a headspace of air were inoculated with diluted fresh activated sludge from a municipal wastewater treatment plant in Helsinki (suspended solids 4 mg l-1). The test was run for 28 d at 20°C with continuous shaking. The biodegradation or mineralization to carbon dioxide was determined by measuring weekly the net increase of total inorganic carbon (TIC) levels over time compared with inoculum or the tested compound in unsupplemented blanks. The inorganic carbon concentrations (TIC) were calculated on the basis of CO2 concentrations in the headspace after acidification to pH<3 with concentrated (>85 % w/v) H3PO4. At this pH the equilibrium constant for the distribution of CO2 gas and liquid phases in the vessels was 1.0, and hence only the IC concentration in the headspace needed to be analyzed. The extent of biodegradation was expressed as a percentage of the theoretical amount of inorganic carbon based on the amount of the initial addition of the tested compound.

Shake flask batch test

The biodegradation of 14C-radio-labelled reference chemicals was tested with a shake flask batch test ISO/DIS 14592-1 (2000) at different concentration regimes (0.002–26 mg carbon l-1 corresponding to 0.004–25.9 mg l-1 as chemical) with different inocula.

Testing was carried out in three replicate rubber septum sealed 1l infusion bottles (exact volume 1220 ml) with a liquid volume of 300 ml and a 920 ml headspace. The 14C- labelled test substances were added to obtain a final activity of 50 000 Bq l-1 and made up to the tested concentration with unlabelled analytical grade chemical. The tested radio- labelled substances were aniline-C-14-UL 250 Ci/ 0.53 mg and from an other batch aniline-C-14- UL 250 Ci/ 0.6 mg, and 4- chloroaniline-C-14-UL 250 Ci/ 1.5 mg and 4- chloroaniline-C-14-UL 250 Ci/ 3.3 mg. The bottles were incubated at 20°C in near dark or diffuse light with continuous shaking. The maximum test duration was 65 days for 4- chlorophenol and samples were taken weekly. The biodegradation was monitored by recording the decrease in total organic 14C activity (TOA) by liquid scintillation counting of acidified and air stripped samples. This TOA consists of residual test compound and its metabolites, accumulated microbial biomass and cell fragments. The total activity (TA) is then equal to the sum of above mentioned TOA and inorganic activity (IA), which was stripped off as CO2 in acidic solution. The degradation rate constant (k) was estimated assuming first order kinetics as the slope of a logarithmic plot of residual activities as a function of time. The half-life (t 1/2) was calculated as t 1/2 = ln(2) k-1.

For the mass balance estimations and for overall check of the test system the activities of different fractions (gas phase, total activity, dissolved residual activity and particulate activity) were also measured by scintillation counting in different trap and scintillation cocktails.

Calculation of Results and Data Analysis

Effluent toxicity (Papers I and II) was generally expressed as Toxicity Units (TU). This unit was based on the EC50 dilution at which the effect in the toxicity test was 50 % compared to controls (TU = 1/EC50 x 100).

In order to evaluate the correlation between toxicity and wastewater chemistry, an overall toxicity index (TI) was computed on the basis of the most sensitive tests (luminescent bacteria, algae and zebra fish egg and sac fry test, Verta et al. 1996). The original toxicity results of each wastewater or effluent sample in each tests were classified to 5 toxicity classes (TC). Those samples which were not toxic in the test in question had a toxicity class of 0. For the others, values from 1 to 4 were given according to their position in the frequency distribution of the test in question. If the sample toxicity lay within the first fractile it was given a value of 1, if in second fractile, value 2, etc. Finally the toxicity index (TI) was calculated according to the formula:

$$TI = (TC_L + TC_A + TC_{ZH} + TC_{ZS})/16, (1)$$

where T_{CL} = toxicity class according to the luminescent bacteria test, TC_A = toxicity class in the algae test, TC_{ZH} = toxicity class according to zebra fish hatching and TC_{ZS} = toxicity class based on zebra fish survival. Effects of sample waters on microbial processes (Paper II) and biomass in the mesocosm studies (water column bacterial counts, sediment biomass and sediment monophosphoesterase activity data) were tested with two way (time and sample water treatment) analysis of variance (ANOVA) with log (x+1) transformation and also with the non-parametric Kruskal-Wallis test because of the distribution of the data. The possible differences observed in ANOVA were tested by the t-test and correlations by direct Pearson and by linear regression. The two control mesocosms had to be considered separately because of the different light and temperature conditions.

Soil toxicity (Papers III and IV) was generally expressed as Soil Toxicity Units (STU) or EC50 values of diluted soil elutriates (g soil l-1). The STU was based on the EC50 dilution of soil at which the effect in the toxicity test was 50 % compared to controls (STU = 1/EC50 x 1000).

In pesticide studies the measured field data (Paper V) of soil microbial activity and biomass of six replicate study fields of each pesticide or cultivation treatment on six sampling occasions, separately during the two growing seasons was evaluated by ANOVA (SAS Institute Inc. 1988). The ANOVA was performed for each sampling of each treatment and if any possible significance was found at a risk level of 5 % or lower ($p < 0.05$), pair-wise t – tests were performed. Additionally, the possible effects of other environmental factors (soil water content, organic carbon content, NO3 - and NH4 + concentrations) were also statistically evaluated by stepwise least square regression analysis (LSA) (SAS Institute Inc. 1988).

In the study of biodegradation in different conditions (Paper VI), the biodegradation rate and half-life were calculated for the reference chemicals. The degradation rate constant (k) was estimated assuming first order kinetics as the slope of a logarithmic plot of residual activities as a function of time. The half life ($t_{1/2}$) was then calculated as $t_{1/2} = \ln(2) \cdot k^{-1}$.

RESULTS AND DISCUSSION

Assessment of Environmental Effects of Industrial Effluents

The toxicities of different bleaching wastewaters and treated effluents from the two pulp mills determined by the Vibrio fischeri luminescent bacteria, Pseudomonas putida, algae and Daphnia tests as Toxicity Units (TU). The ECF and TCF bleaching wastewaters before secondary treatment from Mill A1 strongly inhibited the bioluminesce in the V. fischeri test. These wastewaters were even more toxic than the tested conventional chlorine bleaching wastewater. The untreated mixed hardwood and softwood bleaching wastewaters from Mill B1 were less toxic than the wastewaters from MillA1. The secondary treated effluents did not exhibit any toxicity to luminescent bacteria.

The bacterial growth inhibition test with P. putida showed only slight toxic responses to untreated chlorine bleaching and to one ECF (birch) bleaching wastewater. Untreated TCF wastewater and all secondary and treated effluents stimulated the growth of the bacteria. Only the untreated bleaching wastewaters inhibited the algal growth. There were no clear differences in the wastewater toxicity between conventional bleaching and ECF or TCF bleaching. One of the untreated ECF (birch) wastewaters was more toxic than the chlorine bleaching wastewater. Most of the secondary treated effluents stimulated the algal growth. The Daphnia were less sensitive than the luminescent bacteria and algae. However, in the same study the hatching and survival of the fertilized Zebra fish embryos and sac fry stage larvae sometimes appeared to be the most sensitive toxicity test.

The toxicities of different wastewater streams (92 samples) from the mills, representing debarking, black liquor condensates, ECF and TCF bleaching sequences (alkaline, acid and total), untreated total effluents and secondary treated effluents were assessed by the luminescent bacteria and the mitochondrial particle tests (Paper I). The results are expressed as Toxicity Units (TU). The most toxic wastewater seemed to be the debarking process waters and black liquor condensates when assessed with both screening tests. The TCF wastewaters were very toxic in the luminescent bacteria tests. Unfortunately, the number of TCF bleaching effluents assessed in this study was too limited ($n = 3$) to make valid statements concerning their average toxicity. After the secondary treatment the effluents gave only sporadic and slight toxic responses. Although the data are not shown here, it is worth mentioning that freezing and one month of storage at –18°C did not have a statistically significant effect (t-test, $p > 0.05$) on the toxicity of the samples (Ahtiainen et al. 2002).

In the first phase of the study a toxicity index (TI) was derived on the basis of luminescent bacteria, algae and zebra fish test results (Paper I and Verta et al. 1996). This overall toxicity based on the four most sensitive measured endpoints representing different trophic levels (bacteria, algae, fish) appeared to correlate better with sample organic carbon content than with AOX. This rather poor correlation of biological effects and halogenated organic compounds such as AOX has also been documented in an earlier study (Hodson, 1996). For example mixed function oxygenase (MFO) induction in fish has been often strongly associated with pulp mill effluents. However, chlorine is not an essential component of these MFO inducing compounds and thus AOX concentrations in pulp mill effluents and surface waters are unrelated to MFO induction potency. Verta et al. (1996) also found some correlation of overall toxicity with the concentrations of total phenols, fatty acids and dichloromethane (DCM) extracts. However, they could only partly explain the predicted toxicity by using phenol, fatty acid and resin acid concentrations in their regression model.

The correlation of the TOC content and luminescent bacteria toxicity was not obvious in the screening study. This could be explained partly by some highly toxic samples from TCF bleaching, which might have contained some peroxide residuals. Unfortunately these could not be chemically analysed. The eukaryotic mitochondrial particle test and luminescent bacteria test were poorly intercorrelated ($r = 0.266$, data not shown).

The results indicate that the natural substances in the wood, and their transformation and degradation products, could be the bioactive compounds in pulp mill wastewater. Especially wood debarking waters and black liqueur condensates, which contain high amounts of these mostly unknown substances (Axegård et al., 1993), appeared to be acutely toxic. The toxicity of TCF and ECF bleaching wastewater appeared to be a result of many bioactive substances rather than a few specific compounds. These and other compounds from the pulp bleaching may be eliminated by suitable biological secondary treatment. This may be the primary reason for the observed reduction in effluent toxicity.

For the above-mentioned reasons it is very difficult to predict the environmental impact of novel pulp mill effluents on the aquatic biota by using only defined chemical analysis. Laboratory toxicity tests will continue to play an important role in directing management decisions regarding the environmental effects of pulp mill process modifications (Kovacs and Megraw 1996). Until we know more about the possibly harmful substances originating from wood, the use of bioassays in the assessing of pulp mill effluents is recommended. A simple set of biotests such as the algae,

Daphnia and luminescent bacteria assays could be used. The luminescent bacteria test appears to be the best for preliminary screening and classifying effluent toxicity. Furthermore by combining the bioassays and effluent fractioning, some of the problematic compound groups could possibly be identified and the processes could be improved. This kind of toxicity identification and evaluation (TIE) approach can be applied to industrial effluents or to any complex mixture (Kszos et al. 1992, Ankley and Burkhard 1992). Furthermore, a similar strategy (toxicity reduction evaluation, TRE) can be used to guide and improve the process and the wastewater treatment for effective removal of toxicity (USEPA 1988).

Toxic loads from pulp and paper mill effluents to recipients were compared with effluents from chemical, metal and textile industries (previously unpublished data). The toxicity (TU) and the toxic loads of secondary treated effluents are presented in Table 9 as toxicity emission rates (TER). Certain effluents of the metal and textile industries were very toxic especially to algae (highest TU = 50). However, the effluent toxicity and possible environmental effects should always be considered together with the effluent flow and receiving water body. The effluent flow from pulp and paper mills is often great compared to other industries and hence even sporadic toxicity can cause significant effect in the recipient. The sensitivity of different tests varies depending on the toxicant composition in the effluent. As seen in this limited data, effluents from the metal and mining industry (six samples) or textile industry (six samples), possibly containing heavy metals were occasionally especially toxic to algae. The luminescent bacterial test and the algae growth inhibition tests have often been observed to be the most sensitive tests in routine monitoring of the toxicity of industrial effluents (Diehl et al. 1995). This has also been stated by Wängberg et al. (1995) who studied the toxicities of various industrial waste waters. Especially an algal test battery with different phylogenetic groups of microalgae in multiwell plates increased the sensitivity to detect toxicity compared to standard algae tests.

However, some of the bioactive compounds in pulp mill effluents can have potential long term effects (reproduction and community changes) on aquatic biota. These impacts on community structure and function cannot be assessed by short term laboratory toxicity tests. The zebra fish egg and sac-fry tess give some indication of chronic effects but usually model ecosystem studies (mesocosms) or long scale monitoring data are required to assess community changes.

The long term effects of ECF and TCF bleaching wastewaters, secondary treated effluents, natural humus water and treated municipal effluent on the microbial biomass (total cell number) development in the mesocosm studies (Paper II) is presented in Fig. 3. Compared to the variation of the two control mesocosms no drastic or statistically significant effects (two-way ANOVA) of the waste waters was detected. Bacterial counts varied between 2×10^6 and 8×10^6 cells per ml. The untreated wastewaters occasionally increased the biomass in the water column compared to controls during the summer periods. This could be due to the nutrients and dissolved carbon in the wastewaters. However, at the end of the growing seasons no statistically significant ($p > 0.05$) differences between mesocosms were detected.

The active microbial biomass as ATP content in sediment and its seasonal fluctuation in mesocosms exposed to 1/400 dilutions of sample water is presented in Fig. 4. The sediment biomass showed elevated fluctuation compared to controls during the one and a half year experiment in all the mesocosms. Occasionally, the biomass of the two control mesocosms differed, but the difference was not statistically significant (t-test, $p > 0.05$). The difference between controls might be due to the different light and temperature conditions of these mesocosm basins, as seen in an earlier study (Lehtinen et al. 1997).

Alkaline phosphomonoesterase (PME) activities in the surface sediment in mesocosms exposed to 1/400 dilutions of sample water are presented in Fig. 5. There were no overall statistically significant effects ($p > 0,05$) of different waste waters on the enzyme activity. However, the same kind of increased oscillation could be detected in treated mesocosms compared to control mesocosms as in the sediment microbial biomass. Often a relationship between sediment surface phosphatase activity and chlorophyll suggests that benthic algae might be a major source of phosphatases (Chappel and Goulder 1994), or that there exists a link between algal activity and bacterial phosphomonoesterases. It has been suggested that the exudated organic carbon from the photosynthetic algae directly enhances the phosphatase activity in the same sediment biofilm (Espeland and Wetzel 2001).

In this study the PME activity did not decrease during the winter, under ice and snow cover. In some mesocosms (ECF HD, ECF LD) the highest peaks in activity (four times higher than controls) were measured on 5th March under the thick ice cover. On this single sampling occasion the ECF LD and ECF HD mesocosms differed significantly from the controls (ANOVA: $p = 0.02$ and $p < 0.01$ respectively). This may be due to a very early diatom algal bloom or to higher bacterial metabolic activity during the winter. The mesocosm exposed to municipal effluent (MUN) reached the highest enzyme activity in spring (three times higher than controls on 6th May statistically significant by ttest, $p < 0.05$) but the activity collapsed during the summer. At the same time this mesocosm appeared to pass through nutrients, especially nitrogen (Tana et al. 2002). The system lost more ammonium than was introduced and consumed about 80 % more nitrate than controls. This negative ammonium value could be a clear sign of degradation of the biomass produced.

Exposure to the waste waters did not change the qualitative diversity of enzyme activities measured by API Zym strips (data not shown here). Almost the same enzymes gave positive reactions both in the mesocosms exposed to bleaching waste water and in the control mesocosms. Untreated TCF wastewater and naturally humic water gave the highest diversity of positive reactions of different enzymes. A possible explanation for this could be that these waters contained higher amounts of different substrates and carbon sources than were received by the other mesocosms. However, the changes in enzyme profiles between sampling occasions were inconsistent and difficult to interpret. Hence no statistical evaluations could be applied. The inconsistency of the results might be due to the fact that the enzyme activities chosen for this commercial kit originate from clinical microbiology and their relevance to aquatic enzyme processes may be limited. However, the API Zym kit yielded approximately similar results to the quantitative measurements of the PME activity.

The toxic effects of the bleaching wastewaters could be observed in the laboratory tests (Paper I). This toxicity to microbial biota could not be clearly statistically demonstrated in the mesocosm studies (Paper II). The varying environmental conditions and the exposure were together the most important factors affecting the microbial variables in the mesocosms. The methods used assess mainly the changes in total biomass or microbial activity. Hence these measurements will only reveal severe harmful effects, and fail to indicate the possible harmful changes in the structure of the microbial community. The changes in the microbial community structure may result in long term effects on nutrient cycling and on food webs. However, when the two main stress factors caused by waste waters were toxic and eutrophying substances, the effects were sometimes hard to interpret. How it is possible to discriminate between toxic and eutrophying responses? Eutrophication should not be understood as increased growth rates of all biota. A clearly beneficial nutrient or carbon source increase for one species may be detrimental to other organisms.

After the initial characterization of effluent toxicity and possible effects on the biota by a larger set of bioassays representing different trophic levels, some more simple tests can also be applied for routine toxicity monitoring and generally no acute toxicity should be detected in the secondary treated effluents of modern pulp mills. Biotests, and among them microbial tests can provide a useful tool for the environmental hazard evaluation of mixed industrial effluents possibly containing unknown harmful substances. Microbial tests also have an important role in the identification and evaluation of toxic substances and in guiding toxicity reduction by improving industrial or treatment processes. The biological quality surveys in the recipient or in extensive mesocosms studies such as in this study contain much environmental variation and the possible detrimental effects of pollution are difficult to observe. This environmental variation is meant to be partly controlled and exactly documented in mesocosm studies. However, the experimental design (enough replicates to have statistical power) must be compatible with the available resources and with financial realities.

Toxicity Assessment of Polluted Soils

The toxicity of chlorophenol contaminated soil and wood chips decreased during composting as assessed by both modifications of the luminescent bacteria tests (Paper III). This decrease of toxicity followed decreasing chlorophenol concentrations in the compost piles (Table 10). Chlorophenols, as direct un-couplers of oxidative phosphorylation and hence of energy production for luminescence, are usually highly toxic in luminescent bacteria tests. The EC50 concentrations for several chlorophenols are between 2 and 5 mg l-1 (Ribo and Kaiser 1987). In this study about 200 mg total chlorophenols per kg dry weight soil appeared to be the threshold value for chlorophenol toxicity in the tests. The reason for this high threshold value for toxicity, is most probably the limited bioavailability of chlorophenols. Strong adsorption to soil organic matter reduced the toxicity of pentachlorophenol in studies by van Gestel and Ma (1988) and probably also of other chlorophenols. This has also been documented in microcosm studies in which the effects of PCP (50 and 500 mg per kg d wt soil) on soil invertebrate communities and microbial activities were assessed (Salminen et al. 1995). Only the high concentration (500 mg kg-1) reduced microbial activities as measured by soil ATP content and soil respiration.

The toxicity and the total PAH content of soil in a two-year period of bioremediation of soil heavily contaminated with polyaromatic hydrocarbons in the smaller pilot pile composting. In this smaller pilot study the toxicity assessed by the solid phase luminescent bacteria test, the standard luminescent test and the P. putida growth inhibition test did not decrease in the inoculated pile as much as in the control pile. The increase of total and some identified PAHs in the inoculated pile during the first summer composting was also noteworthy. A probable explanation for this was the heterogeneity of the compost. The tarry creosote was unevenly distributed in the pile, which made representative sampling of the pile difficult. However, increased toxicity was not observed in the toxicity tests after the first period, although the PAH concentration increased. After the second summer almost complete degradation of PAHs and similar reduction in toxicity was observed in both the pilot piles.

In the larger scale case study fluctuating but generally decreasing toxicity was also recorded during the five months of composting. The plateau phases in degradation were also detected by all toxicity tests as temporarily increasing toxicity. The explanation could be the increase of individual PAH compounds with higher toxicity, such as benzo(a)pyrene. Certain unknown transformation products, other aromatic hydrocarbons with high toxicity, or degradation to more toxic or more bioavailable individual compounds could also cause increased toxicity. Ultimately the reduction of PAHs was almost 80% and the decrease in toxicity could be seen with both modifications of the luminescent bacteria tests but not so clearly with the Pseudomonas growth inhibition test. The toxicity of the soil to P. putida was rather low throughout the composting. However, the observed oscillation of toxicity based on STU units correlated well with the other tests. The Toxi-Chromo Pad test indicated similar reduction of toxicity to enzyme synthesis in bacteria. After two and a half weeks of composting no inhibition was observed.

The toxicity and chemical data gathered from these case studies on chlorophenol and PAH contaminated soils is rather limited (pooled single samples) and descriptive. Hence strong statistical conclusions cannot be drawn. The microbial toxicity tests, especially the solid phase luminescent bacteria test, proved to be a practical tool for assessing the toxicity during bioremediation of creosote and chlorophenol contaminated soil. Even if the high metal content causes its own basic level of toxicity during composting, it did not mask the toxicity caused by changing concentrations of organic contaminants. The semi-quantitative or more qualitative Toxi-Chromo Pad test also gave indication of reduced toxicity during composting. However, in the final assessment of the environmental risks of contaminated or bioremediated soil, a set of different trophic level biotests should be used (Maxan et al. 2000) and other environmental concerns, such as genotoxicity (Belkin et al. 1994) should also be evaluated.

As in the case of effluent assessments, the biotests have the value of being applicable to guiding the contaminated soil treatment processes, and they should be used together with chemical analysis in the final evaluation of remediated soil. The soil matrix definitely requires the evaluator of the soil quality to more adequately consider many factors such as the contaminant distribution, homogeneity and bioavailability compared to effluent risk evaluation. However, many similar uncertainty factors of sampling and analysis are involved both in chemical analysis of defined compounds and in the determination of the limit concentrations of substances, as in using biological toxicity tests.

Effects of Agrochemicals on Soil Microbes

The different pesticide treatments did not have remarkable effects on soil microbes measured by soil respiration, nitrification and soil ATP content in the field studies. The microbial activity and biomass mainly followed the weather conditions. The study strategy simulated realistically normal crop production processes, and the sampling program, experimental design and analyses of the field samples were planned to assess the possible acute and long term effects of normal pesticide use. However, because of the large temporal and spatial scale of the field experiment the variability of the environmental factors (aerial soil heterogeneity, weather) did not allow strong statistical interpretation.

The observed changes were usually not statistically significant. The only statistically significant difference (ANOVA: $p < 0.02$) was the contemporary elevated nitrification potential after the chlorsulfuron herbicide treatment. Hence it was not possible to conclude that the reduced and supervised pesticide use (AD, BD) was less harmful to soil microbes than the conventional use (AC, BC), as did Jones et al. 1992 in a similar study. On the other hand, it must also be understood that the AC and BC treatments with pesticide use represent the accepted and normal (and not extreme) barley production conditions in Finland.

Many pesticides have been assessed for their side-effects on carbon and nitrogen transformation in soils. Anderson (1992) reported studies in which the effects of 68 different pesticides at normal, fivefold and tenfold doses on soil carbon and nitrogen mineralization processes were assessed. Only 5 out of 68 pesticides caused a decrease of 15 % in nitrification potential at the normal dose. Moreover only 15 of the pesticides studied had some effects even at the highest doses. The pesticides were not named and probably none of them were the same as those used in this survey.

In the laboratory soil respiration inhibition tests, some effects of the propiconazole would have been expected on the basis of toxic responses in the luminescent bacteria test, in which the EC50 value was 1.25 mg l-1. However, the only effective concentration of propiconazole was 125 mg kg-1. During the field experiment propiconazole was found in the surface soil at a maximum concentration of 2 mg kg-1 (data not shown). Hence the lack of clear field effects on soil microbes was reasonable. The fungicide propiconazole selectively affects fungi, and it has been reported that it decreased the length of fungal hyphae in a field test (Emholt 1992). However, agricultural soils tend to have bacterial-based energy channels rather than fungal mechanisms as in forest soils (Moore and Hunt 1988). In one earlier laboratory study, propiconazole also caused doseresponse effects on soil respiration only at concentrations higher than those used in the field (Emholt 1992).

In the field studies and in the laboratory assessments (soil respiration, nitrogen and carbon mineralization processes), a natural heterogenic population of soil microbes was exposed to the chemical for various periods of time. If there were sufficient numbers of resistant species or taxonomic groups in the exposed community, the activity of these microbes could camouflage the possible effects on other, more sensitive microbes. This was not the case with single species tests, which could be seen as more sensitive but which are considered to be ecologically less relevant. In agricultural soils the key microbial processes in nutrient circulation should be assessed. It should also be understood that one intention in cultivation is to control the diversity of the microbial community for more favorable crop production conditions, for example avoiding denitrification and fungal diseases.

The high clay (>70 %) and organic matter (2.6– 4.1 %) contents of the studied soil appeared to have an effect on the bioavailability of the chemicals, hence reducing the exposure of organisms to the pesticides studied. The rapid reduction of soil toxicity to luminescent bacteria could be seen in the bioavailability assays. The rapid toxicity reduction during one day could not be explained by the degradation of the chemicals and it was probably due to rapid adsorption to soil particles and hence reduced bioavailability. This kind of limited bioavailability of dimethoate has also been observed in studies on soil fauna (Martikainen et al. 1998). In the same study dimethoate was also documented to degrade rather rapidly.

The Finnish boreal environmental conditions and the poor bioavailability of chemicals in soil may lead to poor biological degradation of substances and possible accumulation of pesticide residues in soil. Whether this is a risk, especially when a field is turned to forest or pasture (changes in pH, organic carbon content and soil structure), should also be carefully assessed. The overall use of pesticides has mainly been decreasing during the past twenty years in Finland (Hynninen and Blomqvist 2001). Today, new pesticides are very efficient. For example only 4 g of chlorsulfuron herbicide is needed for an area of one hectare. This sets high demands for the chemical analysis of trace amounts of modern pesticides and their residues in soil.

The field assessments of the effects of chemicals on soil microbes can be evaluated by combining the chemistry, biovailability data, soil and substance toxicity and observed biological responses. The study design is very important in field experimental work and the varying environmental conditions can interfere with treatment responses. In the case of a severe contamination detrimental effects on soil microbial biomass and activity have been seen, as in the case of airborne heavy metal contamination measured by similar methods to those used in this study (Vanhala and Ahtiainen 1994). However, the initial risk assessment of pesticides will also mainly be based on the results of harmonized laboratory studies and toxicity tests in future chemicals policy. The field assessments have their value in more definitive ecological research as an effort to combine and understand the effects observed on various parts of the ecosystem food webs.

Biodegradation of Chemicals

The results of this study partly support the theory that low concentrations provide different biodegradation kinetics than concentrations used in the standard tests. The degradation of the studied reference chemical aniline at different concentrations was always rapid in the standard test conditions compared to rather slow degradation of chloroaniline. The high concentration (25.9 mg l-1) of biodegradable aniline in standard test conditions with activated sludge appears to support exponential growth of degraders.

Unfortunately, the biomass was not monitored in this particular experiment by ATP measurements during chloroaniline biodegradation, but it usually remains rather low (data not shown). However, the source of the inoculum appeared to be an even more important factor affecting biodegradation rates than the concentration of the tested chemical. A broad range of biodegradation rates (Table 2 in Paper VI) and half-lives (Fig. 11) was observed for both tested reference substances at different concentrations with various sources of the inocula.

Aniline is well known as a readily biodegradable substance in standard test conditions (Painter 1995) and almost total degradation was also observed observed in a eutrophic lake (Subba-Rao et al. 1982). In this study aniline degraded rapidly in the ISO standard test conditions and in urban sea water, also at low concentrations of 0.002 mg l-1 (half lives 5–8 days). Aniline satisfied the criteria for ready biodegradability reaching 60 % mineralization of the theoretical carbon dioxide production. In activated sludge and in urban harbour sea water the aniline degraders appear to be present in rather high numbers. However, degradation was clearly slower in unpolluted sea water (half lives 10–20 days) and the high concentration of aniline (25.9 mg l-1) yielded a broad range of rather inconsistent half lives of 40–150 days. The higher concentration might already have been close to toxic levels for some microbes. In natural lake waters degradation was generally much slower. In oligotrophic lake water the half lives at low concentrations varied from 35 to 70 days. In a humic lake with high organic carbon content the aniline degraded poorly. The reason could be that the microbes are adapted to degrade only the natural carbon pool present in the lake and not aniline-like compounds.

Chloroaniline degraded very slowly in all natural waters and often the degradation even failed, e.g. in pristine sea water and in humic lake water. The fastest degradation was observed in urban harbour water with half lives between 15 and 120 days. This range was wide and the results were rather inconsistent. Similar inconsistency in biodegradation half-live, with more recalcitrant compounds was earlier observed by Ingerslev and Nyholm (2000). In standard biodegradation test conditions the lag phases have also been long and sometimes the biodegradation has even failed.

The biodegradation rates of the chemicals tested in Finnish natural waters appeared to be lower (near zero especially with 4- chloroaniline) than the rates documented in some other countries (Nyholm and Toräng 1999). An important factor for degradation in natural environments is the number of potentially adapted degrading microbes present. If the discharge of a certain compound is continuous the microbial population will develop the capability to mineralize these molecules. In pristine environments the variety of biodegradation potential for man-made substances is probably more limited. This could clearly be seen in this work when comparing the pristine (Tvärminne) and urban sea (Helsinki harbour) areas. In wastewater treatment the most important factor in the establishment of an adapted population, in addition to continuous discharge of the compound, is the sludge retention time (van Ginkel et al. 1995). This time defines the age of the activated sludge and how often it will be renewed.

The source of the inoculum appeared to be more important factor influencing the rate of biodegradation than the concentration of the tested compounds in this study. However, the data were very limited, and especially in standard test conditions with activated sludge, when the biomass was also assessed by ATP analysis, different growth kinetics of the microbes were observed depending on the tested concentrations. The concentration normally used in the standard tests (20 mgC l-1) caused rapid and exponential growth of microbial biomass. Aniline, a readily degradable substance, degraded well in activated sludge regardless of the concentration. In pristine natural water the difference in degradation rates between 0.002 and 0.022 mg carbon contents of the tested compound per litre could be indicated. 4-Chloroaniline appeared to be such a poorly degradable compound in Finnish conditions that the effect of concentration on the degradation rates could not be clearly established.

Technically, the these modern biodegradation testing methods functioned well. The ISO 14593 headspace carbon evaluation test (Battersby 1997) provides a modern, practical "ready biodegradability" screening method with rather stringent conditions with which volatiles and poorly soluble substances can also be tested. The shake flask batch test with radio-labelled compounds (Nyholm et al. 1996, Ingerslev et al. 1998) can produce data simulating many environmental compartments such as natural surface waters, but also activated sludge at environmentally relevant concentrations.

Biodegradation testing is an important tool in estimating the exposure of biota in various environments to chemicals for the risk assessment. It has great value in conjunction with the chemical analysis and toxicity tests in effluent studies. If an effluent gives a toxic response, due mainly to unknown organic substances, biodegradation information on this mixture of substances could provide guidance to treatment improvements, or help to estimate the exposure of biota in the environment receiving the effluent.

Because of the importance of biodegradation data to exposure estimations in risk assessment schemes, basic research on biodegradation kinetics in different environments is greatly needed. This should direct the harmonized test guideline development towards more environmentally relevant conditions to provide better estimates of biodegradation rate constants.

Applicability of Microbial Tests and Measurements in Environmental Assessments

Microbial tests applied in toxicity assessment of industrial wastewaters, polluted soils and agricultural chemicals yielded useful and relevant information (combined with the chemical analysis) on the possible environmental effects of complex mixtures of harmful substances and chemicals. The industrial wastewaters could be ranked for their toxicity using microbial tests. These tests appeared to be rather simple and convenient tools for toxicity screening of different inmill wastewater streams. The bioremediation of chlorophenol and PAH polluted soils could be monitored and evaluated using the two modifications of the luminescent bacteria test. The potential microbial effects of pesticides used in the field studies could be screened more sensitively using single species toxicity tests than in indigenous microbial community tests such as the respiration inhibition test. The solid phase modification of the luminescent bacteria test appeared to be especially useful in bioavailability assessment of pesticides in soil.

Because of the above-mentioned observations the microbiological toxicity tests could be of use in the toxicity identification and evaluation (TIE) and toxicity reduction and evaluation (TRE) schemes applied in aquatic and terrestrial environments. The microbiological tests also have a role in the regulatory context, such as in pollution permits and in regulatory routine monitoring.

The microbiological single species (or strain) test methods often appeared to be the most sensitive tests compared with tests with other organisms in the wastewater studies. The responses in microbial tests had some correlation with other tests and with the analysed chemical variables, and hence they could be used as surrogate analyses for screening purposes. However, the bacterial and algal tests had their intrinsic value in the set of biotests, as both groups of organisms have important functions in the ecosystems.

Microbiological biomass and activity measurements were applied for evaluating the effects of pulp mill wastewaters in aquatic mesocosm studies, and of pesticides in terrestrial field and microcosm studies, on microbial communities. The used measurements proved to be applicable and usually adequately sensitive to observe the changes, but the environmental variation and limitations in the experimental design did not enable strong statistical conclusions in either of the studies.

The study on biodegradation kinetics with the two reference compounds revealed some important differences between the standard biodegradation test method, routinely used in the biodegradation assessments, and the simulation test method. The rate of biodegradation depends clearly on the source of the inocula used in the tests, and also on the tested concentration of the compound in the test. This should be considered when estimating the half-lives of chemicals in various environments on the basis of the current data provided by the standard tests.

SUMMARY AND CONCLUSIONS

The microbiological toxicity tests proved to be rather good indicators of effluent toxicity. The luminescent bacteria test and algae test were usually the most sensitive tests, was also reported in other studies. The use of biotests together with chemical analysis is to be recommended in evaluating and monitoring complex mixtures (such as pulp and paper mill effluents) containing unknown substances. The toxicity of pulp mill wastewaters correlated well with the total organic carbon content of the samples (TOC, COD, BOD) but less strongly with the AOX content. After secondary treatment, in which most of the carbon pool and toxic substances appeared to be degraded, the effluents gave only sporadic and weak toxic responses. The treated effluents from chemical, metal and textile industries were occasionally toxic. However, the effluent toxicity should always be considered together with the effluent flow and the hydrological characteristics of the receiving water body. The data on the wastewater or effluent toxicity can be used to guide the industrial processes or the wastewater treatment in toxicity reduction evaluation (TRE) schemes. By combining toxicity testing, sample fractioning and manipulation and chemistry the potentially toxic substances could be identified in complex mixtures as in toxicity identification and evaluation (TIE) strategy.

Unfortunately, the use of biotests in routine and regulatory monitoring has been rather limited in Finland. Most of the monitoring and regulatory permits for effluents are based on chemical analyses. The new Water Framework Directive 2000/ 60/ EU sets demands for analysing certain harmful priority substances from the effluents and from the environment and for assessing the ecological quality in the environment. It does not directly support biological testing of effluents, although the biological effects on biota (ecological quality) is the main objective of the directive. Similarly, the Integrated Pollution Prevention and Control (IPPC) directive (96/61/EU) is based on chemical variables.

As in effluent studies, microbial tests (either on soil elutriates or directly in contact with soil) provide data for the final evaluation of remediated contaminated soil. The biotest can also guide the selection of the best available remediation techniques. If the soil is contaminated for example with known heavy metals it is rather safe to evaluate the environmental risk on the basis of chemical analyses. However, if the soil is contaminated by several organic substances and metals, which is often the case, the use of simple in the characterization and mapping of the contamination can be useful. In the case of pure heavy metal contaminations, biotests can be applied in the actual bioavailability testing in order to refine the exposure estimations.

Microbial measurements have an important role in the assessment of biological quality of aquatic or soil environments. Because of the other varying environmental factors such as weather, the minor effects on microbial communities are difficult to prove statistically.

Microbial communities have a strong and sometimes rapid potential for community changes, which can mask the possible effects on sensitive microbes. If even slight significant effects are observed in the level of the whole community biomass or activity, these evidently affect the overall ecosystem functions, as seen in areas with strong heavy metal pollution. One practical option to assess community changes could be the measurement of functional diversity e.g. by enzyme activity profiles.

The effects on microbes should also always be considered in the risk assessment of chemicals. This is already the case for pesticides but should be broadened to include other industrial and household chemicals. In practice, biodegradation data is sometimes used in chemicals risk assessment as source of information on toxicity to microbes.

The biodegradation of chemicals is an important factor affecting the exposure of biota to chemicals. This fact has been noted in the EU technical guidance document for the risk assessment of chemicals. The scientific development in biodegradation testing has been rather slow despite the urgent needs. Often the interest of both industry and regulators has been the desire to keep the obligatory biodegradation testing as cheap and simple as possible. Hence, most of the existing OECD test guidelines have been developed to screen whether the chemical has the potential to be easily degraded or not. These test guidelines are not meant to predict the rate of biodegradation in the environmental compartments. The efforts to create a sound scientific basis for modelling of fate and behaviour of chemicals have often been rejected with the arguments that either the research project would last too long, whereas environmental legislation could not wait, or the results of the project would be too complex for practical use.

However, both basic science and test method development must in the future better meet the requirements of chemicals policy in biodegradation testing. As seen in this study and in other studies earlier, biodegradation of chemicals in the standard tests and in various environments differs sometimes significantly. The reasons for this, in addition to many environmental abiotic factors, are the high concentrations of the tested chemical in the standard tests compared to those in the environment, and the variability of inocula from different sources. Whether we can accept these differences should be considered, and the results observed in this and other studies should direct the testing strategy and test guidelines development towards simulation of more environmentally relevant conditions.

The microbial reactions in toxicity tests and in the environment do have some correlation with the reactions of other biota. In certain cases, especially in the screening or monitoring of the possible changes, for example in effluent quality or in soil remediation processes, they can serve as surrogate tests for toxicity to other biota. However, the microbial reactions have a value of their own. The microbes are needed in the environment to sustain the food webs and in various mineralisation processes.

In most cases environmental risk evaluation requires chemical analysis of the suspected contaminants, toxicity testing of the contamination (chemical, effluent, contaminated soil or sediment) and survey of the biological quality (activity, biomass, diversity). In this contex, when the importance of microbial processes and of the microbial link in the food webs in various environments is established, the microbial toxicity tests and microbial measurements in the environment should be included in the risk assessments of various environmental pollution problems.

8

Bacterial Toxicity Tests versus Chemical Analysis

There is a need for cheap and reliable methods for assessing the efficiency of bioremediation processes.Could often expensive chemical analysis be supported or partly replaced by cheaper microbiological toxicity tests?

The biodegradation of polycyclic aromatic hydrocarbons (PAH's) was followed in two creosote oil contaminated outdoor composts by chemical analysis and by bacterial toxicity tests.Both of the two compost piles were otherwise similarity treated during the composting period but the other pile was inoculated with PAH degrading mycobacteria.Chemical analysis and toxicity assessments were carried out in the beginning and in the end of four month composting period.

The pooled soil samples were homogenized and PAH's were quantified and identified by a chromatographic (GC/MS) method.Also the metal contamination (Cu, Cr and As) was evaluated by AAS method.The toxicity of the soil samples was assessed by the direct contact modification of the standard Vibrio fisheri luminescent bacteria test and the toxicity of water extracts from these soil samples was also evaluated by luminescent bacteria test and by the automated modification of Pseudomonas putida growth inhibition test.

Total amount of PAH's was reduced in both piles and especially the loss of 2 and some 3 ring compounds was almost

90 per cent.However the concentrations of some PAH's increased during composting.Toxicity assessments support clearly chemical data.There was an apparent decrease in toxicity in both piles after the four month period but the inoculated pile containing more 3 and 4 ring compounds was clearly more toxic.The direct contact modification of V. fisheri luminescent bacteria test appears to be a cheap and suitable tool for assessing toxicity changes during bioremediation processes.

IDENTIFICATION OF CONSERVED TRAITS IN FLUORESCENT PSEUDOMONADS WITH ANTIFUNGAL ACTIVITY

A collection of 29 fluorescent pseudomonads, some with known biological control activity against a range of phytopathogenic fungi, were characterized phenotypically and genotypically by comparing carbon source utilization patterns, suppression of *Pythium ultimum* both *in plants* and *in vitro* and the potential to produce known secondary metabolites. Fatty acid profiling and restriction fragment length polymorph-ism (RFLP) analysis of the ribosomal DNA operon (ribotyping) were used to determine the diversity of isolates. A small group of genetically related *Pseudo-monas* spp. with similar properties was identified; each isolate produced a diffusible bioactive product *in vitro* and was active against *Pythium ultimum in planta.* However, other isolates that were able to sup-press damping off disease but did not inhibit hyphal extension *in vitro* clustered outside this group. Pheno-typic analyses revealed that the accumulation of C17:0 cyclopropane fatty acid (17CFA) and the pro-duction of hydrogen cyanide correlated significantly with biological control activity and with the antagon-ism of fungal development. The potential of 17CFA as a marker for the selection of fluorescent pseudo-monads with biocontrol agent (BCA) potential was demonstrated by the isolation of a novel active strain. This was selected after the screening of 13 clonal groups of fluorescent pseudomonads identified from 500 isolates from the phytosphere of sugar beet. Levels of 17CFA synthesis possibly reflect the effi-cacy of the *rpoS* allele in particular strains.

INTRODUCTION

Many fluorescent *Pseudomonas* spp. are capable of reducing the incidence of plant diseases caused by soil-borne fungi. Few studies have examined the taxonomic and functional diversity of these bacteria. However, it is apparent that a large variety of antifungal metabolites are pro-duced by these organisms. Hydrogen cyanide (HCN) (Voi-sard *et al.*, 1989), phenazines (phz) (Thomashow and Weller, 1988), phloroglucinols (phi) (Fenton *et al.*, 1992; Keel *et al.*, 1992), pyoluteorin (pit) (Kraus and Loper, 1995) and pyrrolnitrin (prn) (Pfender *et al.*, 1993) are pro-duced by many pseudomonads, and attempts have been made to quantify the importance of each in disease sup-pression (Osburn *et al.*, 1989; Maurhofer *et al.*, 1994). However, production of characterized bioactive metabo-lites does not account for all of the observed antifungal activity (Kraus and Loper, 1992).

The suppressiveness of soils directly correlates with the density of populations carrying the phi biosynthetic locus (Raaijmakers *et al.*, 1997), which has been detected in phenotypically and genotypically distinct pseudomonad isolates (Keel *et al.*, 1996). However, expression was dependent on the genetic background of the strain into which the phi locus was inserted (Fenton *et* aL,1992), indi-cating that a variety of other factors influence biocontrol efficacy for a given pseudomonad. The high capacity of fluorescent pseudomonads to compete for seed exudates has been implicated in their rapid establishment on roots (Gamliel and Katan, 1992) and may thus lead to the exclu-sion of deleterious fungi. *Pseudomonas putida* N1 R reduces the amount of Pythium-stimulatory seed exudates avail-able to the fungal spores (Paulitz, 1991). The efficient assimilation of iron from soil has also been implicated in biocontrol efficacy (Leong, 1986; Becker and Cook, 1988; Loper and Buyer, 1991) and is attributed to the production of siderophores (Leong, 1986). However, these features alone do not fully explain the known disease suppression attributes of many fluorescent pseudomonads (Hamdan *et al.*, 1991; Paulitz, 1991).

It is therefore highly relevant to establish which features of these fungal antagonistic strains distinguish them from other fluorescent pseudomonads. It is probable that they are all specialized for survival in soil and may be capable of the regulated production of antifungal compounds in the soil-root environment, but do they share other common phenotypes? Some important insight may be gained by

determining the ecological function of secondary metabolite production in bacteria and the environmental pressures that have led to the evolution of these pathways. It may be that production of these metabolites has evolved as a mechan-ism for survival under intense competition (Mazzola *et al.*, 1992). Concomitantly, it has been suggested that those bacteria that demonstrate non-specific plant growth pro-motion may have evolved highly efficient signalling sys-tems with the host plants, which lead to specific plant-microbe interactions (Pierson and Pierson, 1996).

At present, the majority of potential biological control agents have been selected after the screening of large numbers of isolates for their ability to control disease *in planta* (Campbell, 1989). Seedling assays remain one of the most reliable methods for this, as they ensure that any antifungal activity observed is an *in situ* phenomenon. However, they are expensive and time-consuming to run, and the results are often variable. The *in vitro* identification of traits indicative of biocontrol activity would permit screening of larger numbers of isolates at a greatly reduced cost. Thus, the aim of this study was to compare phenotypic and genotypic characteristics of fluorescent pseudomonads with proven biocontrol potential against phytopathogenic fungi and relate the results to the ability to suppress *Pythium ultimum* infection in pea seeds.

RESULTS

Comparative analysis of antifungal activity

The efficacy of each strain for the control of *Pythium ulti-mum* infection of pea seeds is shown (Table 2) and is con-trasted with the *in vitro* inhibition of hyphal extension on agar plates. Three distinct levels of antagonism were observed on plates. First, the bacterial colony was com-pletely overgrown; secondly, *Pythium* grew to the edge of the bacterial colony but not over it; thirdly, a distinct zone of inhibition was seen around the edge of the bacter-ial colony where *Pythium* growth was not observed. No correlation between *in vitro* antagonism and *in situ*disease suppression was observed. For example, F113 did not inhibit *Pythium ultimum* growth on plates, but effectively suppressed damping off [control index (CI) = 0.87], whereas PH6 generated a zone of inhibition on agar with a CI of only 0.33 (Table 2).

Fatty Acid Profiling

Fatty acid profiling of these 29 pseudomonads revealed a considerable degree of divergence within the collection (Fig. 1). By comparison with the commercially available database (MIDI, Newark, DE, USA), all but one of the strains were named within the rRNA homology group I pseudomonads (Table 1). The exception to this was PGSB 8456, which did not match with any entries in the database. An orange pigment, produced in abundance by this strain, was co-extracted with the fatty acid methyl esters (FAMES) and consequently interfered with the gas chromatographic analysis. No correlation was detected between the similarity of total FAME profiles of strains and their ability to suppress disease (Fig. 1). However, with the exception of strain 76/10, isolates with elevated levels of C17:0 cyclopropane fatty acid were effective in the sup-pression of damping off disease in peas. On this basis, the proportion of C17:0 cyclopropane fatty acid in each of the strains was identified as a variable for further correlation and stepwise multiple regression analysis as described below (Table 3).

Ribosomal DNA restriction fragment length polymorphism (RFLP) analysis

Ribotyping, as described here, demonstrated that there was a large degree of genetic variation in this collection of pseudomonads and that the majority of the strains were genetically distinct (Fig. 2). PGS12 and PGSB 8456 could not be differentiated by this method. In addi-tion, both strains also had a very distinctive orange pig-mentation when grown on solid media. Two other closely related strains, CHAO and Pf-5, also had similar FAME nrnfiles (Fig. 1)_

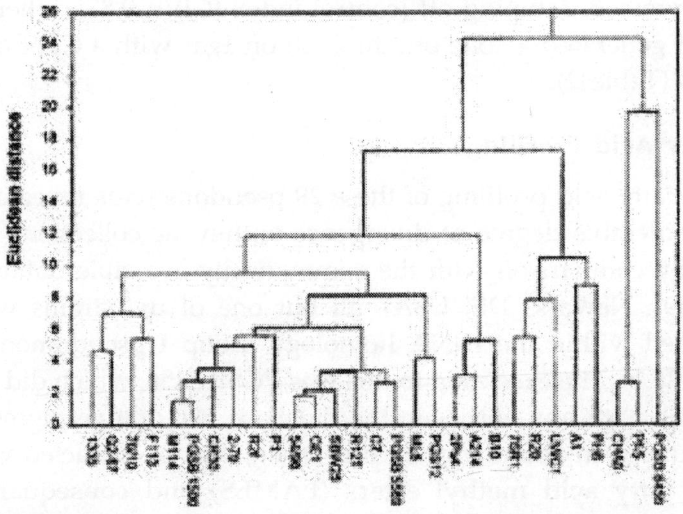

Fig. 1. Dendrogram indicating the relatedness of the biological control strains listed in Table 1 according to the analysis of whole-cell fatty acids. Euclidean distances are calculated on the basis of the relative proportions of individual fatty acids of pairs of strains.

Although there was no direct correlation between taxo-nomic relatedness and absolute biocontrol activity, a small cluster of strains was identified that exhibited similar disease-suppressive and *in vitro* antagonistic activities (Fig. 2).

Table 1. *Pseudomonas* strains used in this study.

Isolate	Source	Habitat	MIDI ID[a]
SBW25	Bailey *et al.* (1995)	Sugar beet	*P. fluorescens A*
54/96	Zeneca Agrochemicals	phylloplane, Oxford	*P. fluorescens A*
1335	Zeneca Agrochemicals	Sugar beet, Belgium	*P. chlororaphis*
7611C	Zeneca Agrochemicals	Sugar beet, Belgium	*P. chlororaphis*
2-79	Thomashow and Weller (1988)	Sugar beet, Belgium	*P. marginalis*
Q2-87	Bangera and Thomashow (1996)	Wheat, Washington,	*P. fluorescens C*
CHAO	Voisard *et al.* (1989)	USA	*P. putida A*
Pf-5	Kraus and Loper (1992)	Wheat, Washington,	*P. putida A*
2Ps4	van Elsas	USA	*P. cichorii*
GE1	van Elsas	Tobacco, Switzerland	*P. fluorescens A*
P1	van Elsas	Cotton rhizosphere,	*P. chlororaphis*
R2f	van Overbeek and van Elsas (1995)	Texas, USA	*P. putida A*
R12T	van Elsas	Soil, Netherlands	*P. fluorescens A*
PH6	Fuhrmann and Wollum (1989)	Soil, Netherlands	*P. putida A*
C7	Latour *et al.* (1996)	Grass rhizosphere,	*P. savastanoi*
M114	Fenton *et al.* (1992)	Netherlands	*P. fluorescens* B
F113	Fenton *et al.* (1992)	Grass rhizosphere,	*P. savastanoi*
Al	Fukui *et al.* (1994)	Netherlands	*P. putida A*
CR30	Fukui *et al.* (1994)	Grass rhizosphere,	*P. marginalis*
ML5	Osburn *et al.* (1989)	Netherlands	*P. chlororaphis*
PGS12	Georgakopoulos *et al.* (1994)	Soybean rhizosphere,	*P. chlororaphis*
R20	Osburn *et al.* (1989)	NC, USA	*P. savastanoi*
UWC1	Cardiff, UK	Flax rhizosphere,	*P. putida A*
PGSB1500	Plant Genetic Systems	France	*P. fluorescens* B
PGSB5589	Plant Genetic Systems	Soil, Ireland	*P. savastanoi*
PGSB8456	Plant Genetic Systems	Soil, Ireland	No match
7SR1	Buyer and Leong (1986)	Potato periderm, CA,	*P. aeruginosa*
A214	Buyer and Leong (1986)	USA	*P. viridiflava*
B10	Buyer and Leong (1986)	Field soil, CA, USA	*P. cichorii*
		Beet spermosphere,	
		CA, USA	
		-[b]	
		Bean rhizosphere, CA,	
		USA	
		-	
		Sugar beet	
		rhizosphere, Belgium	
		Sugar beet	
		rhizosphere, Belgium	
		Sugar beet	
		rhizosphere, Belgium	
		-	
		-	
		-	

a. Closest match in MIDI Aerobe library v3.8' as *determined* by *FAME analysis.*
b. Information not known.

These pseudomonads had a control index greater than 0.7 and produced an antifungal compound that could diffuse through agar, as identified by a zone of clearing around the colony. These strains were greater than 63.8 ± 6.8% genetically similar, as determined by their RFLP ribotype. Only two strains within this cluster, 2Ps4 and 1335, did not exhibit this combination of antifungal activity. Both showed some activity on plates, although there was no zone of clearing for 2Ps4, but performed poorly in disease suppression (CI <0.5).

Phenotypic analysis of isolates

BIOLOG carbon source utilization patterns further demon-strated the metabolic diversity of isolates (Fig. 3). Strains used between 33 (PH6) and 60 (1310 and PGSB 5589) of the 95 different carbon sources available, and no correlation was found between related clusters and antifungal activity.

The metabolic potential, substrate utilization patterns for each isolate and the phytopathogen *Pythium ultimum* were determined and expressed as the nutritional similar-ity index (NSI). The values obtained for isolates ranged between 0.560 (strain PH6) and 0.857 (strain 54/96) (Table 2). The majority of strains (> 85%) exhibited an NSI of greater than 0.7, demonstrating considerable simi-larity in the substrates used by *Pythium ultimum* and the fluorescent pseudomonads. Considerable differences in the growth rate of isolates in pea seed exudate were observed (Table 2). Mean generation times (MGTs) varied between 2.227 h (strain UWC1) and 6.596 h (strain Q2-87) with the MGT for most isolates (> 90%) less than 4 h.

Siderophore production, as determined by the colour change of chrome azurol S in an agar medium, was also highly variable (Table 2). Although some strains (C7) appeared not to produce any compounds with affinity for ferric iron, other isolates produced prolific amounts, creat-ing a large zone of orange (up to 180 mm^2, i.e. UWC1).

The potential of each strain to produce well-characterized antifungal metabolites was determined. Cyanide and phenazine-1-carboxylate production was assessed by screening bacteria for the products *in vitro*. Genotypic analyses were performed for the presence of genes homologous to 2,4-diacetylphloroglucinol or pyoluteorin biosynthetic loci (Table 2). Neither of these methods were able to determine whether biosynthesis of the com-pounds occurred *in situ*. Over 62% of the isolates in the collection showed a positive response to any of the tests, despite their observed ability to control damping off in pea seedlings.

Correlation and regression analysis of data

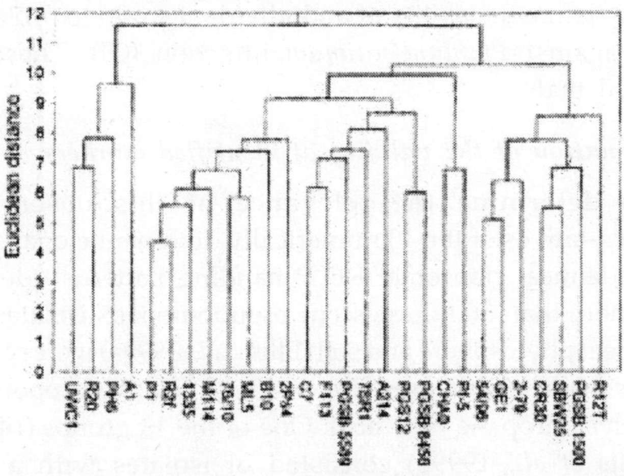

Fig. 3. Dendrogram indicating relatedness between the strains listed in Table 1 as determined by BIOLOG metabolic profiling. Euclidean distances are calculated on the basis of the differences between profiles. These data were then clustered using the unweighted pair group method of averages.

The matrix indicating the correlations between all charac-teristics analysed is shown in Table 3. The correlation the only variable that could be used rigorously to predict activity was the ability to produce cyanide. However, the strong correlation of C17:0 cyclopropane fatty acid with both CI and HCN production implied that FAME analysis could serve as a highly suitable marker for the primary screening of collections to identify potentially effective *Pseudomonas* biological control agents. between cyanide production and the proportion of C17:0 cyclopropane fatty acid was the highest at 0.7524 ($P<0.001$). Only these two traits showed significant corre-lations with the control index.

Small but significant negative correlations were found between siderophore production and mean generation time in pea seed exudate (PSE) and siderophore produc-tion versus NSI. A small positive correlation existed between NSI and MGT. All comparative data were sub-jected to stepwise multiple regression to test whether a combination of factors could be used to define the ability of an individual strain to protect pea seeds against *Pythium ultimum* infection (CI). These tests revealed that

Confirmation of the validity of identified markers

To determine the relevance of this observation, represen-tatives of the 13 numerically dominant clonal groups (with identical ribosomal RFLP banding patterns) identified from more than 500 fluorescent pseudomonads isolated from sugar beet phytosphere and soil (Ellis *et al.*, 1999a) were screened for the characteristic presence of a high relative proportion of C17:0 cyclopropane fatty acid. One of the 13 groups (ribotype 11; Ellis *et al.*, 1999a) consisted of isolates with a C17:0 cyclopropane fatty acid content of 15%, whereas the other groups consisted of isolates with a C17:0 cyclopropane fatty acid content of less than 5%.

A single isolate from the ribotype 11 clonal group, *P. chlororaphis* S34/10, was assayed for the ability to sup-press *Pythium ultimum* infection. The control index of S34/ 10 (CI ˆ 0.83) was equivalent to those strains that had been selected for biocontrol efficacy by more traditional and protracted (seedling assay) methods. Therefore, FAME analysis of strains from a large collection of natural isolates facilitated the direct and successful identification of an effective antifungal biological control agent.

Discussion

Despite the observation that the collection of *Pseudomo*-nas strains shared the common feature of being able to suppress *Pythium ultimum* infection of pea seeds *in planta,* isolates represented a genetically and phenotypi-cally diverse group. FAME profiling identified the majority of the strains as members of the rRNA homology group I pseudomonads (Table 1), as described by Palleroni (1984). In some cases, the identification provided by FAME profiling did not correlate with established affilia-tions; this is a result of the inadequacies of the taxonomic libraries for closely related strains rather than variability in the reproducibility of the FAME profiles of isolates.

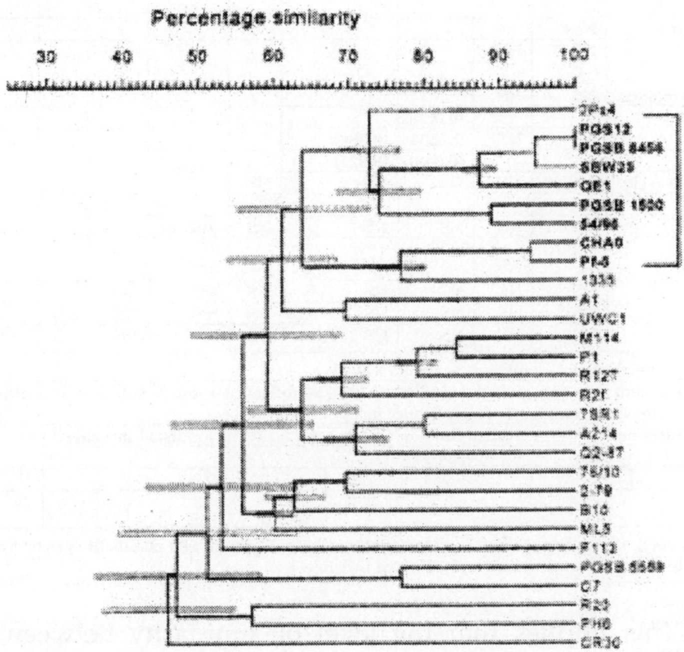

Fig. 2. Dendrogram indicating genetic relatedness between the strains listed in Table 1 as determined by RFLP analysis of the rDNA operon on two enzyme digests of genomic DNA. The scale shows percentage similarity as determined by Dice coefficients.

However, analysis of these and other characteristics facilitated assessment of the extent of inter-relatedness and the importance of a previously unreported FAME marker for antifungal activity. Versatility and variation is typical of *Pseudomonas,* but it was anticipated that isolates from plant-associated habitats would exhibit some traits in com-mon. The limited genetic similarity between the known BCA strains (Fig. 2) was similar to that seen in other studies using similar methods for the assessment of relatedness of fluorescent pseudomonad populations (Rainey *et al.,* 1994; Ellis *et al.,* 1999a).

Table 3. Spearman rank correlation coefficients for the traits listed in Table 2.

	CI	*In vitro*	MGT	Siderophore	NSI	C17:0 cyclo	HCN	phz	phl	plt
CI	-									
In vitro	NS[a]	-								
MGT	NS	NS -	-	-						
Siderophore	NS	NS	0.3751							
			(0.025)							
NSI	NS	NS	0.3304	-0.3862	-					
			(0.043)	(0.021)						
C17:0 cyclo	0.4330	NS	NS	NS	NS	-				
	(0.011)									
HCN	0.5497	NS	NS	NS	NS	0.7524	-			
	(0.001)					(< 0.001)				
phz	NS	NS	NS	NS	NS	NS	NS	-		
phl	NS	NS	NS	NS	NS	NS	NS	NS	-	
plt	NS	NS	NS	NS	NS	NS	NS	NS	NS	-

a. Correlation not significant at the 95% level.
Coefficients are only given where P<0.05 (95% significance level). Numbers in parentheses indicate the actual level of significance as determined by the one-tailed Student's Rest. Keys for traits are listed in Table 2.

Table 2. Disease suppression, nutritional, antifungal, growth and metabolite production potentials of *Pseudomonas* strains listed in Table 1.

This implies that the level of simi-larity between the predominantly BCA strains in this col-lection is no greater than that of any random group of *Pseudomonas* spp. selected from soil or plant tissue samples. Thus, antifungal activity appears to be a general and genetically widely dispersed function of soil and phy-tosphere pseudomonads and not a trait specific to a specialized, genetically restricted group of pseudomonads.

Cluster analysis based on phenotypic traits (FAME and carbon source utilization (Figs 1 and 3 respectively) demonstrated the phenotypic diversity of the collection. However, the lack of congruence between clusters from phenotypic (Figs 1 and 3) and genotypic (Fig. 2) data was interpreted as a reflection of functional and genetic diversity of plant growth-promoting bacteria (Rainey *et al.*, 1994; Natsch *et al.*, 1997). Neither of the methods applied provides a definitive classification of the strains and, therefore, care should be taken in interpreting data to reveal inter-relationships that are based on a single method of analysis.

Niche overlap indices, as determined by nutritional pro-filing (NSI), have been used to identify strains that are cap-able of co-existence in the same habitat (Janisiewicz, 1996) and have been reversed to identify organisms that are not compatible (Wilson and Lindow, 1994). However, nutritional profiles of fungi and bacteria have not been compared previously for the purpose of biological control. The ability of pseudomonads to use a wide range of car-bon sources has been well documented (Palleroni, 1992; Schroth *et al.*, 1992) and is reflected in these results, but the nutritional versatility of *Pythium ultimum* was unex-pected. The relatively high similarity between the meta-bolic potentials of the fungi and bacteria may reflect the range of nutrients available in soil and plant-associated habitats. The problem with comparing metabolic profiles of distantly related organisms is that BIOLOG plates pro-vide no information on the relative affinity of each organ-ism for any given substrate and, therefore, the relevance to the ability to use carbon sources at ecological concen-trations is speculative. However, the profiles do identify the potential for metabolism and thus compare the poten-tial for competition, although the substrates found in BIOLOG plates may be of limited environmental relevance. A posi-tive correlation between NSI and MGT values (Table 3) supports the view that nutrient competition plays an impor-tant role for individual isolates in the control of fungal pathogens, although no correlation was drawn between either NSI or MGT and the control index for the group as a whole.

The lack of correlation between *in vitro* inhibition of fungi and the ability to suppress disease caused by those fungi *in planta* has been documented previously (Reddy *et al.*, 1993). This may be attributed to the differential expression of genes *in situ* and *in vitro.* As the regulation and expres-sion of bacterial genes will be different under different conditions, such as artificial media compared with environ-mental growth conditions, it is axiomatic that all potential biocontrol agents must be screened for their *in planta* activity. As observed above, the potential to produce anti-fungal metabolites is of relevance, but competitiveness and ecological fitness are also significant traits. Forexample, hydrogen cyanide production correlated with disease sup-pression (Table 2) and, in previous investigations, HCN has been identified as an important factor in biocontrol activity (Voisard *et al.*, 1989), in which the distribution of cyanide production was estimated to be - 50% in some collections of pseudomonads (Bakker and Schippers, 1987; Nielsen *et al.*, 1998). Of the 29 isolates examined in detail here, nine produced HCN under laboratory con-ditions. This was a greater proportion than was identified to produce or have the ability to produce any other sec-ondary metabolite, but the genetic diversity of this collec-tion taken from many sources was greater than those used in previous comparative studies. As with other antifungal secondary metabolites, HCN production is regulated under the control of stationary-phase sigma factors (RpoS) and the global activator (GacA) (Reimmann *et al.*, 1997). The ecological function of secondary metabolites is unclear, but they appear to enhance survival. As cyanide increases exudation of nutrients from plant tissue (Astrom,

1991), strains that produce cyanide may be more adept at niche exploitation with increased competitive fitness in the sper-mosphere and phytosphere.

During the course of these comparisons, the novel cor-relation between C17:0 cyclopropane fatty acid (CFA) and antifungal activity was observed. There is no evidence that indicates that the fatty acid has antifungal activity, merely that it serves as a useful phenotypic marker. As with HCN, and other antifungal metabolites, the formation of C17:0 CFA occurs primarily in the stationary phase of the growth cycle. In *Escherichia coli,* CFA synthase is pri-marily under the control of *rpoS* (Wang and Cronan, 1994; Grogan and Cronan, 1997; Eichel *et al.*, 1999); this may also be the case for *Pseudomonas* species. Thus, the level of CFA in any given strain may indicate the overall efficiency of the production of metabolites controlled by stationary-phase regulators, such as the sigma factor U^s. The importance of the complex signalling cascade, including RpoS, has been demonstrated for secondary metabolites in *Pseudomonas aeruginosa* (Latifi *et al.*, 1996). It has been shown that different alleles of *rpoS* exist in *E. coli* (Ferreria *et al.*, 1999) and that they affect the downstream expression of RpoS-controlled genes (Wang and Cronan, 1994). Thus, an efficient production of CFA in a pseudomonad strain, assuming that it is RpoS controlled, may indicate that the strain possesses one of the more efficient *rpoS* alleles. Therefore, other RpoS-mediated genes may also be similarly highly expressed, including those coding for secondary metabo-lites with antifungal properties.

The utility of C17:0 CFA as a marker for the effective suppression of damping off disease in pea was demon-strated by the selection of *Pseudomonas chloroaphis* S34/10 from a collection of sugar beet isolates (Ellis *et al.*, 1999a). C17:0 CFA is common to many fluorescent pseudomonads (Vancanneyt *et al.*, 1996), and all but one strain in this collection had detectable quantities of C17:0 CFA (Table 2). Therefore, screening collections or isolates for elevated synthesis of C17:0 CFA could be of general relevance for the primary screening of biocontrol agents, particularly as the assessment is rapid and does not depend on any prior knowledge of secondary metabolite synthesis.

Although alternative sigma factors are known to affect virulence in P. *aeruginosa* (Suh *et al.*, 1999) and fungal disease suppression in P. *fluorescens* (Sarniguet *et al.*, 1995), it is possible that each of the secondary metabolites identified in these and other strains serves only to indicate other unresolved functions in these pseudomonads. This may explain why the products so far identified do not com-pletely account for the observed biological activity.

Further work is required to elucidate the importance of the *rpoS* genotype and related stationary-phase regula-tors in the expression of bioactive secondary metabolites and the associated diversity in CFA production. A full understanding of the regulation of these metabolites, which are produced as bacterial cells responding to envir-onmental signals that induce the general stress response, will significantly aid the search for effective biocontrol agents.

EXPERIMENTAL PROCEDURES

Bacterial strains: storage and culture conditions

The fluorescent pseudomonad strains and other bacteria and fungal strains used are given in Tables 1 and 4 respectively. *Pseudomonas* isolates were cultured on Pseudomonas agar base supplemented with 10 mg 1-′ cetrimide, 10 mg 1-′ fucidin and 50mgl-′ cephaloridine (PSA-CFC; Unipath) at 28°C.

Inocula for all assays were produced by growing isolates in LB at 28°C with shaking at 180 r.p.m. for 18 h. Cells were then pelleted by centrifugation of the cultures at 5000g for 10 min followed by washing twice in sterile double-distilled water

(SDDW) before resuspending in the original culture volume of SDDW.

Propagation and maintenance of fungi

Pythium ulfimum was stored in the dark at 15°C as infested soil prepared by drying and sieving spent soil from infected seed trials (Murray, 1994). The pathogen was isolated from infested soil by plating diluted soil suspension onto potato dextrose agar (PDA; Unipath) supplemented with 320 mg 1^{-} ′aureomycin (Cyanamid; PDAA; Thompson et at, 1993a) or tap water agar [TWA; 1.2% agar no. 3 (Unipath) in tap water]. Plates were incubated at 20°C for 18 h. *Pythium* was identified by its rapid growth rate (Stanghellini and Hancock, 1970).

Oospore suspensions were produced as described pre-viously (Ellis et at, 1999b). Oospores were finally resus-pended in SDDW and stored at 4°C.

In planta quantification of disease suppression

An assay for studying the factors involved in the biocontrol of *Pythium* was adopted (Ellis et at, 1999b). Bacteria were added by applying a suspension in water, prepared as described above, directly to the soil.

Twenty-five peas were pressed into the surface of the soil, and the assays were incubated in the dark at 21°C to optimize the development of disease. Triplicate plates were set up for all treatments, including uninoculated and mock-inoculated controls, and scored for disease after 7 days. The control index (CI) was calculated as described previously (Ellis et at, 1999b) and is essentially the relative proportion of infected seeds com-pared with the mock-inoculated controls.

In vitro screening of fungal inhibition

Each bacterial strain was grown on PSA-CFC overnight at 28°C. Purified oospores (100 µ1; $^{-1}$ x 10^{8} propagules ml^{-}) were spread on the surface of fresh PSA-CFC (low-iron) plates and allowed to dry. A single colony of each strain was then picked from fresh overnight plates and spotted onto the Pythium-inoculated plates.

After incubation at 28°C for 30 h, the extent of inhibition of mycelial development was assessed. Bacterial colonies overgrown by hyphae scored zero, hyphae at the edge of the colony -1 and a distinct zone of clearing around the colony - 2.

Table 4. Organisms and plasmids used in this study.

Organism or plasmid	Relevant characteristic	Source
Fungi	Causal agent of damping off	Zeneca Agrochemicals
Pythium ultimum	disease	Nickerson Seeds, UK
Plants	Forage pea; susceptible to	Housiaux *et aL* (1988)
Pisum sativum var.	*Pythium* damping off	Fenton *et aL* (1992)
'Bohartyr'	16-23S rRNA *(Pseudomonas*	J. E. Loper (personal
Plasmids (probes used)	*aeruginosa* PAO)	communication)
pAC10	phl biosynthetic locus from F113	Ellis *et al.* (1999a)
pCU203	plt biosynthetic locus from Pf-5	
pJEL5786	> 10% C17:0 CFA	
Other pseudomonads		
P. chlororaphis S34/10		

Fatty acid profiling

Phenotypic differences were assessed by fatty acid methyl ester (FAME) analysis. Isolates were streaked on tryptic soy broth (Difco) with 1.2% (w/v) agar (TSBA) in triplicate and incubated at 28°C for 24 h. Cells (50 mg wet weight) were har-vested into glass test tubes and the FAMEs extracted as described previously (Thompson et at, 1993b).

Ribosomal RFLP analysis

Genomic DNA was isolated from bacteria by an adaptation of the CTAB method described previously (Ellis et at, 1999a). Approximately 1 µg of DNA was digested with either of the restriction enzymes Kpnl or EcoRl (10 units; Boehringer Man-nheim) at 37°C overnight. DNA fragments were separated by field inversion gel electrophoresis (FIGE; Bio-Rad).

Electro-phoresis and transfer of DNA to membranes was carried out as described previously (Ellis et at, 1999a). Membranes were probed with the 4.5kb BamHI fragment from pAC10 carrying both the 16S and the 23S ribosomal RNA genes (Housiaux et at, 1988). The fragment was purified from agar-ose, and 50 ng was labelled with the non-radioactive ECL system as described by the manufacturers (Amersham). Hybridization and visualization of the genes were carried out according to the manufacturers' instructions.

Metabolic profiling

Metabolic profiles of all pseudomonad strains were assessed using the BIOLOG GN (Hayward) microplate system accord-ing to the manufacturer's instructions. After 24 h at 28°C, each well was scored on a four-point scale (0, 1, 2 and 3) according to the extent of colour production observed. The data were used to construct a dendrogram of the similarity of the meta-bolic profiles.

BiOLOG plates were also used to determine the metabolic profile of *Pythium ultimum.* The oospore suspension (- 1 x 10^8 propagulesml$^-$) was used to inoculate the microplates (150 µl per well). Colour development was scored as above after 24 h at 28°C.

Relative growth rates in an 'ecological' substrate

The growth rates of each of the strains were compared in a medium consisting of exudate from pea seeds. Pea seed exu-date (PSE) was produced by adding SDDW (50ml) to pea seeds (50g) and shaking at 20°C for 1 h. The exudate was then filtered through Whatman No. 1 paper before sterilizing by passing through a 0.22 µm filter. PSE was stored for up to 1 week at 4°C. Each strain was grown overnight in LB (10 ml) at 28°C with agitation.

The cells were washed three times with an equal volume of distilled water before resuspen-sion in the same. Suspensions

(10µl) were inoculated into PSE (300µI) in individual wells (n=3) of honeycomb well plates. These plates were placed in an automated microplate reader (Bioscreen C; Lifesciences) and incubated at 28°C with agitation. The optical density at 600 nm wavelength was recorded every 15 min over a 24 h period.

Distribution of the ability to produce known antifungal metabolites

Each strain was tested directly for the production of hydrogen cyanide, phenazines and siderophores. Hydrogen cyanide production was assessed as described elsewhere (Voisard et at, 1989). Briefly, each strain was grown overnight in 200_1il of LB in a microtitre plate.

The wells were covered with indicator paper (Whatman 3MM soaked in 5 mg ml⁻′ cop-per ethylacetoacetate and 5 mg ml⁻′ 4,4-methylene-bis-N,N-dimethylalanine in chloroform and air dried), and the plates were incubated at 28°C. Production of cyanide caused the indicator to turn blue.

All pseudomonads in the collection were assessed for the production of phenazine-l-carboxylic acid as described by Thomashow and Weller (1988). Isolates were grown over-night on TSBA before being streaked onto PDA (Unipath) and incubated at 28°C for 4 days. Dark green pigmentation or crystalline deposits in the centre of colonies was indicative of phenazine production.

Siderophore production was quantified using chrome azurol S (CAS) medium (Schwyn and Neilands, 1987). Glycerol (5% v/v) was used as the sole carbon source, and pseudo-monad selectivity was provided by the addition of 10 mg I⁻′ cetrimide, 10 mg 1⁻′ fucidin and 50 mg 1⁻′ cephaloridine (CFC; Unipath).

When the iron (111) is removed from the chrome azurol S complex by high-affinity siderophores, its colour changes from blue to orange. The diameter of the orange halo around colonies after incubation at 28°C for 2 days was indicative of the relative level of siderophore production of each strain.

The presence of genes encoding phi and pit biosynthesis was determined by hybridization of the membranes described above with genes cloned from strains producing these metabolites (F113 and Pf-5 respectively; Table 4). The entire plasmids were labelled and used as probes. The stringency level was - 60% for all probes. One or more hybridizing bands in the genomic digests indicated the presence of DNA homologous to each of the probes.

Data analysis and statistics

Data analysis was performed using UNisTAT 4.5 for Windows, unless stated otherwise. The mean FAME profile from the replicates was used to construct a library. A dendrogram based on the Euclidean distance between pairs of strains and clustered using the unweighted pair group method with arithmetic averages (uPGMA) was produced using Mis software (MIDI, DE, USA). Significant differences in the proportions of individual fatty acids were determined by analysis of variance (Fry, 1993).

Images of hybridized membranes were captured using the Ge1Doc system (Bio-Rad), and images were normalized to the relative positions of the DNA markers. The genetic related-ness between the strains was assessed by cluster analysis, using uPGMA, of Dice coefficients calculated pairwise between strains (Rainey et al., 1994).

Using the BIOLOG metabolic profiles, the nutritional simi-larity index (NSI) between *Pythium* and each strain was cal-culated by an adaptation of the niche overlap index (Wilson and Lindow, 1994), but included those carbon sources used only by either of the two organisms according to the following equation:

$$NSI = 2n_t/(n_{bca} + n_{py}) \quad (1)$$

where n_b is the number of carbon sources used by both organ-isms, and n_{bca} and n_{py} are the total number of carbon sources used by the potential biocontrol agent and *Pythium* respectively.

The mean generation time of each strain was calculated for the period of exponential growth in the pea seed exudate. All data sets were subjected to pairwise correlations to test for links between the different characteristics. All data were ranked and the Spearman rank correlation coefficient (Zar, 1984) calculated. The significance of each value was calcu-lated by means of the one-tailed Student's t test (Zar, 1984). Stepwise multiple regression (Zar, 1984) was per-formed to determine the importance of each of the factors assessed in this study to the ability to suppress *Pythium*-induced damping off in peas.

9

Role of Hydrosulfide Ions in Methylmercury Resistance

Methylmercury-resistant mutants were obtained from Saccharomyces cerevisiae. They were divided into two complementation groups, met2 (homoserine O-acetyltransferase deficiency) and met15 (enzyme deficiency unknown), as reported previously. It was found that met15 was allelic to met17 (O-acetylserine and O-acetylhomoserine sulfhydrylase deficiency). Methylmercury toxicity was counteracted by exogenously added HS^-, and both met2 and met17 (met15) mutants overproduced H_2S. On the basis of these results, we conclude that met2 and met17 (met15) cause accumulation of hydrosulfide ions in the cell and that the increased level of hydrosulfide is responsible for detoxification of methylmercury.

INTRODUCTION

Mercury is one of the environmental pollutants of most concern because it is highly toxic to living organisms and widely used in industry (4, 6, 10). In studying mercury toxicity, knowledge of the effects of mercury at cellular and subcellular levels is indispensable. For such studies use of model organisms is of importance, and Saccharomyces cerevisiae is appropriate in this respect because it is a unicellular eucaryotic organism, readily manipulated genetically, and easily analyzed biochemically. Ono et al. (13) showed that Hg^{2+} resistance mutations were dominant and were divided into two groups by their linkage relationships. It was also shown that the cell wall acted as an adsorption filter for Hg^{2+} and that the Hg2+-resistant mutants had the cell wall with increased Hg^{2+}- binding capacity. On the other hand, Singh and Sherman (20, 21) isolated methylmercury-resistant mutants and found that the mutants were methionine dependent. The responsible mutations were recessive and were divided into two complementation groups. Although one was identified as met2, which confers deficiency of homoserine O-acetyltransferase (EC 2.3.1.31) (3), the other did not correspond to any of the met mutations known at that time and was named metiS. Whether the met2 and metiS mutations confer methylmercury resistance by the same mechanism remains unknown. To elucidate the biochemical nature of metiS and the mechanism of methylmercury resistance, we isolated methylmercury- resistant mutants and characterized them genetically and biochemically. Here, we present the results of our work.

MATERIALS AND METHODS

Yeast strains and growth conditions

The yeast strains used in this study are listed in Table 1. Standard yeast growth media were used (11, 18). YPD medium contained 2% glucose, 2% peptone, and 1% yeast extract. The synthetic minimal (SD) medium was described by Wickerham (22). For solid media, 2% agar was added. The growth temperature was 30°C.

Genetic procedures

Methylmercury-resistant mutants were isolated as described by Singh and Sherman (20, 21) with minor modifications. Cells were spread on SD agar plates, and a filter paper disc (5 mm in diameter) was placed at the center of each plate. Methylmercuric chloride (5 to 10 μl of a 12 mM solution) was applied onto the discs, and then the plates were incubated for 5 to 7 days. Complementation tests were achieved by masS mating. Strains to be tested were mixed with tester strains on YPD agar medium. The plates were incubated overnight, and the cells were transferred to selective plates (SD medium). The plates were incubated for 4 days, and then the growth was scored.

Growth kinetics

An aliquot of the overnight culture was added to fresh YPD liquid medium; the initial cell density was adjusted to approximately 10^4 cells ml-1. Methylmercury, NaSH, and/or homoserine was added to the growth medium at the desired concentration. Growth was monitored by measuring the increase of the optical density at 600 nm with an automatic cell growth analyzer (Bioscreen C; Labsystems Japan Co., Ltd., Tokyo, Japan).

Measurement of H_2S production

Cells were inoculated on LA agar medium (0.3% peptone, 0.5% yeast extract, 4% glucose, 0.02% ammonium sulfate, 0.1% lead acetate, and 2% agar) and were incubated for a week at 30°C. 'qn this plate, H2S-producing colonies develop dark-brown color due to formation of PbS. A more quantitative analysis was made as follows. Cells were grown overnight in YPD liquid mediumn, harvested, washed twice with water, and then suspended in' Sp liquid medium (360 ml) containing 30 ,uM methionine. N2 gas (5 ml min- 1) was passed through the culture vessel and then through a trap containing 3.75 ml of 1% zipc acetate and 0.125 ml of 12% sodium hydroxide (5). The cqltlire was placed in a rotary shaker (30 rpm). After cutlvation, trapped H2S was assayed by the method of Siegel (1). H2S production (micromoles per cell per hour) was deduced by dividing the amount of trapped H2S by the cumulative cell number and the duration of cultivation.

RESULTS

Mutant isolation. Methylmercury-resistant mutants were isolated as described in Materials and Methods. 4 clear zone of growth inhibition was formed around the filter paper disc containing methylmercury, but there were a few colonies within this region. They were picked, subcloned, and tested for methylmercury resistance by the filter disc method. We obtained five independent methylmercury-resistant mutants (MMR1 through MMR5) in this way. Resistance levels of the mutants were quantitated by growth in YPD liquid medium containing various concentrations of methylmercury. A typical result is shown in Fig. 1.

While a mutant (MMR1) and the parental strain (IS66-4C) responded similarly to methylmercury up to 0.5 μM, the latter grew more slowly than the former at higher concentrations (1.5 to 5.0 μM); at 5.0 μM, the parental strain did not grow at all while the mutant grew up, even though the initiation of growth was somewhat delayed.

Complementation of methionine dependence

The five mutants were subjected to complementation tests using strains containing the defined methionine mutations: met] through met8, metlO, metl3 through metiS, and metl7 (Table 2). Two mutants, MMR1 and MMR2, did not complement strains containing metiS or metl7. This result indicates either that the mutants contain both metiS and metl 7 or that meti5 and metl 7 are alleles of the same gene. Since authentic metiS and metl7 strains did not complement each other (Table 2), we conclude that the latter is the case. We hereafter refer to the mutation as metl7 or metl7 (met15) because the enzymatic deficiency has been defined under the name of metl7; metl 7 causes deficiency of a single protein which acts as O-acetylserine sulfhydrylase (EC 4.2.99.8) and O-acetylhomoserine sulfhydrylase (EC 4.2.99.10) (8, 9, 23). Though mutant MMR3 complemented all strains tested, its complementation with the metl7 strain was distinguishably weak. Since met25, which is shown to be allelic to metl 7 by the gene-cloning studies (2, 17), has been reported to complement metl 7 (8), we assume that MMR3 contains a met25- like mutation; unfortunately, a complementation test using an authentic met25 strain was not conducted because such a strain was not available to us. The two remaining mutants, MMR4 and MMR5, did not complement the met2 strain, suggesting that they contain mutations allelic to met2.

Effect of HS- on methylmercury toxicity. From the S-amino acid biosynthetic pathway of S. cerevisiae (15), we suspected that met2 and metl7 cause accumulation of homoserine and HS- and that either or both of these compounds reduce the toxicity of methylmercury. We therefore examined the effect of these compounds added exogenously. Strain IS66-4C was inoculated in YPD liquid medium containing 5.0 μM methylmercury and various concentrations of homoserine or NaSH, and the growth was monitored (Fig. 2). NaSH dose-dependently counteracted methylmercury toxicity up to 0.1 μM. The antagonistic effect of NaSH saturated or tended to decrease at higher concentrations. We attribute this to the toxicity of NaSH itself, because NaSH markedly inhibited growth (data not shown). In contrast, homoserine (up to 540 μM) did not affect the toxicity of methylmercury (data not shown). The results clearly indicate that HS- but not homoserine counteracts the toxicity of methylmercury.

Production of H2S by methylmercury-resistant mutants. Methylmercury-resistant mutants MMR1 (metl 7) and MMR4 (met2) and their parental strain (IS66-4C) were examined for H2S production; a meti mutant (SF115-lA) defective for sulfate reduction (7) was included as a negative control (Fig. 3). While MMR1 and MMR4 yielded dark-brown colonies on LA agar medium, IS66-4C and SF115-lA did not. It was further found that MMR1 and MMR4 produced substantial amounts of H2S while IA66-4C and SF115-lA did not.

DISCUSSION

We have shown in this study that both met2 and meti 7 confer methylmercury resistance via increased accumulation of H2S. As far as we tested, the methylmercury-resistant mutants were only slightly resistant to Hg2+, and the Hg2+-resistant mutants (13) were not resistant to methylmercury at a detectable level (data not shown). Therefore, we conclude that methylmercury and Hg2+ resistances are different. Hg2+ reacts effectively with sulfhydryl (-SH) groups (1), and presumably because of this property it massively binds to the cell wall and the cell membrane (13) and inhibits their functions, such as transport. Tyrosine transport (14) and cysteine transport (12) are severely inhibited by Hg2+. Since Hg2+ hardly penetrates into the cell (13), the cell must have developed in the course of evolution a protective aid on the cell surface (i.e., cell wall). In contrast, methylmercury because of its hydrophobicity appears to penetrate readily into the cell and causes damage in the cell.

Although exogenously added HS- reduced the toxicity of Hg2+, the strains carrying met2 or meti 7 were not markedly resistant to Hg2+ (data not shown). Thus, we conclude that methylmercury is mainly detoxified by an increased level of HS- in the cell. Although we have presented evidence that the cell detoxifies methylmercury by using HS-, we still do not know what chemical reaction contributes most in this phenomenon. Since it is known that H2S aids the volatilization of methylmercury (16), removal of methylmercury from the cell via such a mechanism is possible. However, it is also possible that methylmercury is converted to Hg2+ in the cell and then reacts with HS- to yield highly insoluble and less-toxic HgS. Studies of the chemical transformation of mercury in the cell and the involvement of HS- in it are needed for a better understanding of the molecular basis of methylmercury resistance.

Bibliography

Blackburn, J.W.: *Environmental Biotechnology for Waste Treatment*, Plenum Press, New York, 1991.

David B.: *Life at a Small Scale: the Behavior of Microbes*, New York, Scientific American Library, 1996.

Ellen, Jo: *Medical Microbiology: a Short Course*, New York, Wiley-Liss, 1994.

Fausta, G.: *Biological Factors in Deterioration of Paper*, Rome, ICCROM, 1985.

Gibbons, N.E.: *Bergey's Manual of Determinative Bacteriology*, Baltimore, Williams & Wilkins Co., 1974.

Kenneth, Frank: *Ecology of Soil-bourne Plant Pathogens, Prelude to Biological Control*, Berkeley, University of California Press, 1965.

Leslie A. Lewis.: *A Novel Approach to the Evolution of Microorganisms*, New York, Vantage Press, 1987.

Michael J.: *Microbiology*, New York, McGraw-Hill, 1958.

Phillip E. Greeson: *Methods for Collection and Analysis of Aquatic Biological and Microbiological Samples*, Washington, D.C., 1989.

Rene, J.: *The World of Rene Dubos: a Collection from His Writings*, New York: H. Holt, 1990.

Steven C.: *Biotechnology, Microbes and the Environment*, San Francisco, CA: Center for Science Information, 1990.

Wood, E.J. Ferguson: *Microbiology of Oceans and Estuaries*, Amstedam, Elsevier Pub. Co., 1967.

Index

♦♦♦